Time Series: Applications to Finance with R and S-Plus

Time Series: Applications to Finance with R and S-Plus

Editor
Johanna Adison

scitus
academics

Time Series: Applications to Finance with R and S-Plus
Edited by **Johanna Adison**

ISBN: 978-1-68117-653-6
Library of Congress Control Number: 2016946520

SCITUS Academics LLC,
616, Corporate Way, Suite 2, 4766,
Valley Cottage, NY 10989

www.scitusacademics.com

Notice

Printed in United States of America on Acid Free Paper

Preface

The study of time series is concerned with time correlation structures. It has diverse applications ranging from oceanography to finance. The celebrated CAPM model and the stochastic volatility model are examples of financial models that contain a time series component. Time series analysis can be useful to see how a given asset, security or economic variable changes over time or how it changes compared to other variables over the same time period. The Financial Time Series applications provide a convenient interface for creating, managing, and manipulating financial time series objects. In the past few years there have been several changes in the financial landscape as well as developments in using time series techniques for financial modeling.

Time Series Applications to Finance with R and S-Plus aims to highlight several of these standard as well as non-standard techniques applied in finance using S-Plus and R as statistical analysis tools. The book covers practical aspects of these models including estimation and testing of the models and shows practical examples. This book is designed to help readers grasp the conceptual underpinnings of time series modeling in order to gain a deeper understanding of the ever-changing dynamics of the financial world. It covers theory and application equally for readers from both financial and mathematical backgrounds.

Table of Contents

CHAPTER 1

ARIMA: An Applied Time Series Forecasting Model for the Bovespa Stock Index

Paulo Rotela Junior, Fernando Luiz Riêra Salomon, Edson de Oliveira Pamplona

Institute of Production Engineering and Management, Federal University of Itajuba, Itajuba, Brazil

ABSTRACT

Due to the relative uncertainty involved with the variables which affect financial market behavior, forecasting future variations in a time series of the Brazilian stock market Index (Ibovespa) can be considered a difficult task. This article aims to evaluate the performance of the model ARIMA for time series forecasting of Ibovespa. The research method utilized was mathematical modeling and followed the Box-Jenkins method. In order to compare results with other smoothing models, the parameter of evaluation MAPE (Mean Absolute Percentage Error) was used. The results showed that the model utilized obtained lower MAPE values, thus indicating greater suitability. This therefore demonstrates that the ARIMA model can be used for time-series indices related to stock market index forecasting.

KEYWORDS

Forecasting, ARIMA, Time Series, MAPE, Ibovespa

1. INTRODUCTION

Economic crises in recent decades and the consequent financial losses demonstrate that markets, financial institutions and investors urgently needed to improve their models to measure and predict the risks to which they were exposed. Equity investments become a great alternative when compared to other applications, especially in long periods [1].

Predicting the future behavior of a time series of data on the Bovespa Index (Index of shares of the São Paulo Stock Exchange) is not an easy task, given the uncertainties related to the variables that affect the behavior of financial markets and how they will impact prices practiced in the future. Studies applied to forecasting financial time series of assets, indices and investment portfolios are an important tool used for decision making in many areas, investment managers, asset pricing, and the areas responsible for risk management.

The genius of the work of Markowitz [2] in proving that the risk of a portfolio consisting of n assets is less than the weighted sum of the risks of each asset, given a different correlation of 1, laid the foundations of portfolio selection with a focus not only profitability, but the relationship between the risk taken and their return. This study, along with the work of Sharpe [3] and Fama [4] originated Modern Financial Theory, which combines the risk taken to the expected return in an efficient relationship.

This paper aims to evaluate the performance of the ARIMA model to predict the time series of the Bovespa Index, measured by

MAPE (mean absolute error percentage) and compare it with other models. Historical data of monthly Bovespa quotations from January 1995 to January 2013 were used. The models were used to compare Single Exponential Smoothing and Double Exponential Smoothing.

This paper is organized as follows: In Section 2, a review of forecasting is presented. In Section 3, the research method is shown. In Section 4, analysis of historical data of the Bovespa Index, data transformation, the necessary adjustments and calculation of MAPE values are all examined based on the results found. Section 5 contains the conclusions of the study.

2. FORECASTING

There are two main types of approaches to demand forecasting: qualitative methods and quantitative methods. The combination of qualitative and quantitative methods is approaching the ideal time to make a good forecast demand [5].

The main qualitative methods are: Panel data approach, Delphi method, scenario planning, educated guess, executive committee consensus, sales force survey, Historical Analysis and Market Research [5]-[8].

Quantitative methods are based on historical data (time series) and assume that past results are relevant for predicting the future [7]. The classical methods of time series are: Moving Average, Exponential Adjustment, Linear Trend, and Nonlinear Trend. These methods require that the series is stationary, i.e. the mean and covariance are constant between the periods. In this context, the auto regressive methods for stationary series, and AR (Auto Regressive) and ARMA (Auto Regressive Move Average) are the most suitable because they generate a more reliable prediction [9].

Another critical effect on time series is the presence of seasonality, i.e. oscillations or disturbances in series occurring at regular intervals of less than one year. And, according to Bacci [8], quantitative ARIMA models can describe two classes of processes: linear stationary processes and linear homogeneous non-stationary processes. The stationary linear processes use basically three types of models: auto regressive model of order p$^{(AR(p))}$; moving averages of order q $^{(MA(q))}$ and auto regressive and moving average process of order p and q$^{(ARMA(p,q))}$. Non-homogeneous stationary linear processes assume that the series are non-stationary in level and/or slope [10]. According to Pindyck and Rubinfeld [11], the amount of times that the original series has to be differentiated to result in a stationary series is called order of homogeneity.

Some stationary random processes (forward constant average over time) can be modeled by means of a mixed autoregressive process and moving average ARMA$^{(p,q)}$. Makridakis, Wheelwrigt and Hyndman [12] state that in this case, depending on the values of p and q, Y_t will depend on the past p values of Y and past q values of Θ errors. This process is Equation (1):

$$Y_t = \varphi_1 Y_{t-1} + \varphi_2 Y_{t-2} + \cdots + \varphi_p Y_{t-p} + \delta + \varepsilon_t - \Theta_1 \varepsilon_{t-1} - \Theta_1 \varepsilon_{t-2} - \cdots - \Theta_q \varepsilon_{t-q} \quad (1)$$

For the process shown in (1), the stationary sum $\varnothing_1 + \varnothing_2 + \cdots + \varnothing_p$ should be less than 1 [11]. According to Pindyck and Rubinfeld [11] and Fava [13], non-stationary series can be transformed into stationary series where observations are differentiated one or more times.

The first differentiation of the data is in Equation (2):

$$Z_t = \Delta Y_t = Y_t - Y_{t-1}. \quad (2)$$

Being:

Y_t = observation Y, in period t of the series Y_t without differentiation;

Y_{t-1} = observation Y, in period t − 1 of the series Y_t without differentiation;

$\Delta Y_t = Z_t$ = observation Z, in period t, belonging to series Z_t with data from the series $Y_t$differentiated for the first time.

The data series will be differentiated for the first time in the following manner: the value of the second datum is decreased from the first; the third will be decreased in the second, the fourth from the third, and so on. With this process, the differentiated series for the first time, Z_t will have one less observation (n − 1 observations) than the original series Y_t.

The second differentiation of data can be represented by Equation (3):

$$W_t = \Delta^2 Y_t = \Delta(\Delta Y_t) = \Delta(Y_t - Y_{t-1}) \quad (3)$$

According to Bacci [8], the series Y_t differentiated a second time, or series Z_t diffentiated once, will lead to the series W_t.

The differential data series Z_t is obtained as follows: the value of the second observation decreased from the first observation forms the first observation, the value of the third diminished from the second provides the second and so on.

The twice differentiated series (W_t) will have n − 2 observations compared to the original series Y_t. Thus, after one or more differentiations of Y_t series to make it stationary, it produces a series stationary W_t, which can now be modeled as an ARMA process(p,q).

According to Pindyck and Rubinfeld [11], the initial series Y_t is an autoregressive process of order (p,d,q) integrated moving average (MA), given by Equation (4):

$$\phi(B)\Delta^d Y_t = \delta + \theta(B)\varepsilon_t \quad (4)$$

Being:

$\Delta^d Y_t = W_t$; with d = order of the stationary series W_t, that is, the number of times that the non-stationary series Y_t was differentiated until becoming a stationary series W_t;

$\phi(B) =$ autoregressive operator;

$\theta(B) =$ moving average operator;

$\Delta = 1 - B$, where operator B imposes a time delay of one period each time it is applied to a variable Y_t.

The construction of an ARIMA model is based on a cycle with the following stages [10]: Identification of a general class of models which will be analyzed; Specification of the model, based on autocorrelation analysis, partial autocorrelations and other criteria; Estimation of model parameters; Verification of the adjusted model, which is done by means of residual analysis to measure its suitability to carry out the forecast; If the model is not suitable, the cycle repeats from the identification of the model.

3. MATERIALS AND METHODS

According to Bertrand and Fransoo [14], this investigation can be classified as applied, having a descriptive empirical goal, since the researchers are interested in creating a model that adequately describes the causal relationships which may exist in reality, leading to the understanding of the current processes thus, fostering understanding of real processes. The approach to the problem was quantitative, and a mathematical modeling research method adopted. Figure 1 shows the Box-Jenkins method [15]:

Initially data collection was done using Economatica® software, using monthly closing prices of the Bovespa Index, for the period January 2000 to December 2012.

4. RESULTS ANALYSIS

The values obtained were plotted using Minitab® 16 Statistical Software for an initial evaluation of the data as shown in Figure 2.

It can be seen that the data is not stationary and the series presents variance from one period to another. The analysis used in this series demonstrated the need for a logarithmic transformation on the data which generate Figure 3. Figure 4 and Figure 5 show ACF (autocorrelation) and PACF (partial autocorrelation) transformed time series.

Tests for ACF (autocorrelation) and PACF (partial autocorrelation) indicated that the AR1 model ARIMA (0, 2, 1) model could be used to predict the behavior of the series, shown in Table 1 andTable 2.

A verification of the series' residuals transformed by Log Bovespa Index and both tests were carried out, through which it was demonstrated that autocorrelation does not exist between series residuals, which enabled the utilization of both to forecast the behavior of the series, as shown in Figures 6-9.

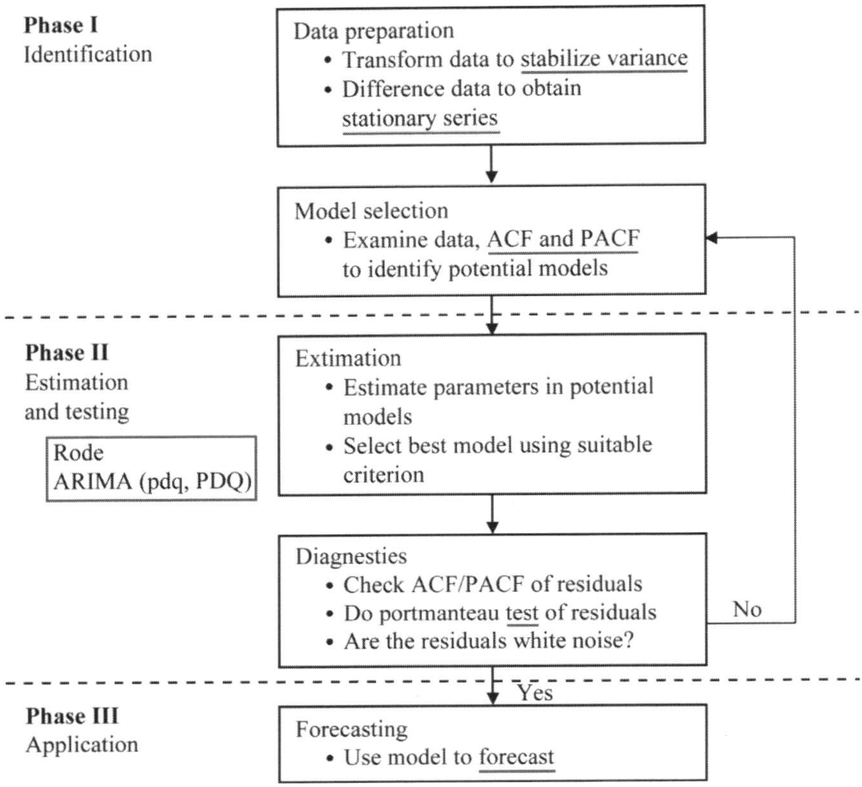

Figure 1. Box-Jenkins Methodology. Source: Makridakis, Wheelwrigt and McGee [15].

The next step was to realize tests to verify the accuracy of the models. Initially the model was used to forecast 10 months ahead in several periods of the series, trying to compare the MAPE between these periods, as shown in Figure 10 and Figure 11.

The use of this indicator (MAPE) to evaluate the models was used because this measures the absolute average, i.e., the sum of the percentage errors, in which the series data values undergo alterations throughout time, which influences the size of the error. Using the value of the error divided by the value of the observation, transforming the error as a percentage of this observation, diminishes the effects caused by variation of the values of the series, allowing one to compare the error between observations of distinct values.

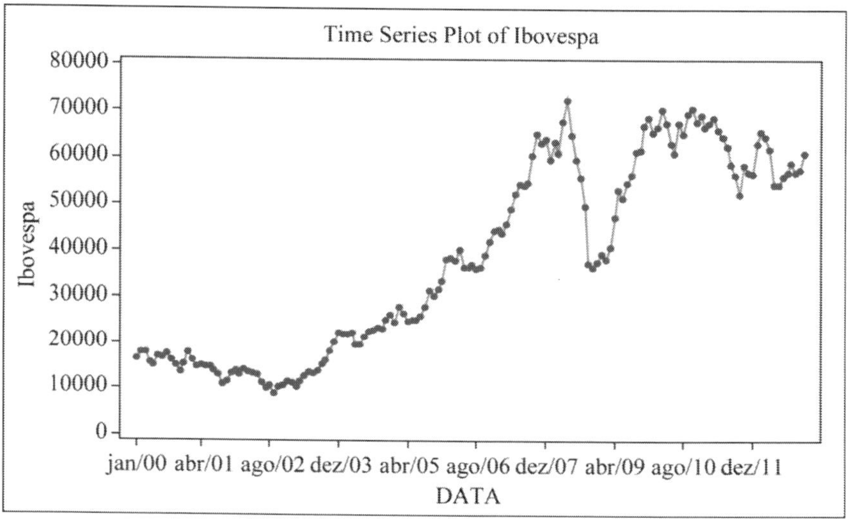

Figure 2. Plotting the Ibovespa time series.

Table 1. Coefficient statistics of the AR1 model.

	Coefficient	Coefficient standard error	T statistic	P-value
AR 1	0.9996	0.0006	1697.01	0.000

Table 2. Chi-square statistics for the Box-Pierce modification (Lhung-Box).

Lag	12	24	36	48
Chi-square	11.5	18.5	26.2	33.5
Degrees of freedom	11	23	35	47
P-value	0.399	0.728	0.858	0.931

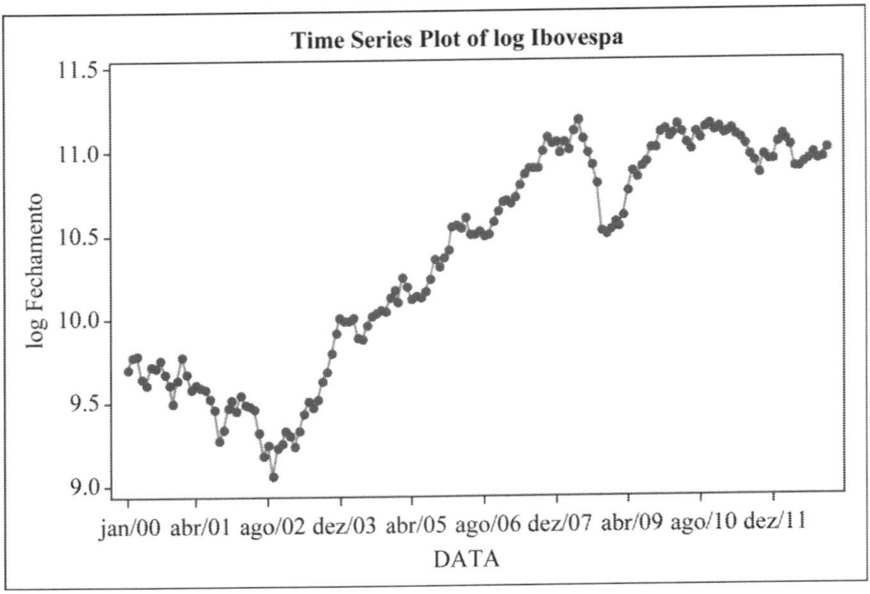

Figure 3. Logarithmic plotting of the Ibovespa time series.

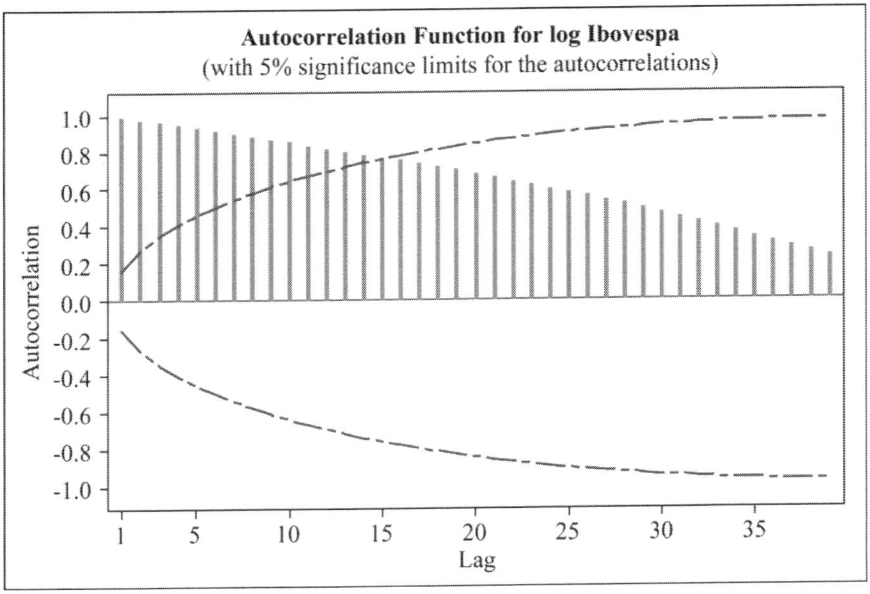

Figure 4. ACF of logarithmic Ibovespa.

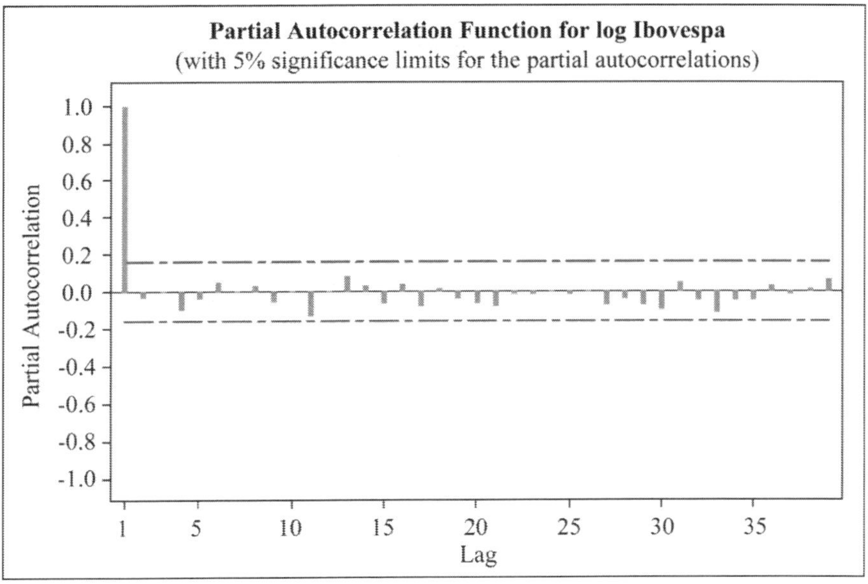

Figure 5. PACF of logarithmic Ibovespa.

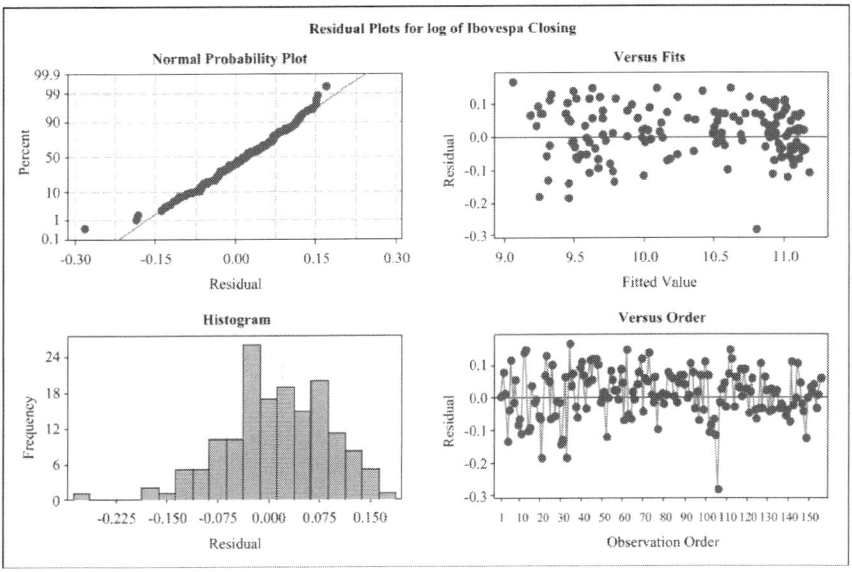

Figure 6. Residual plotting.

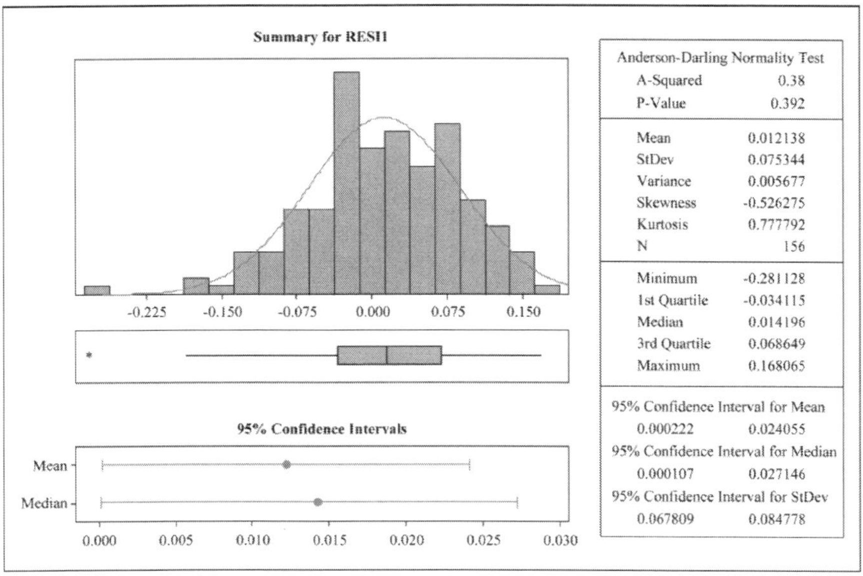

Figure 7. Graphical summary of the residues.

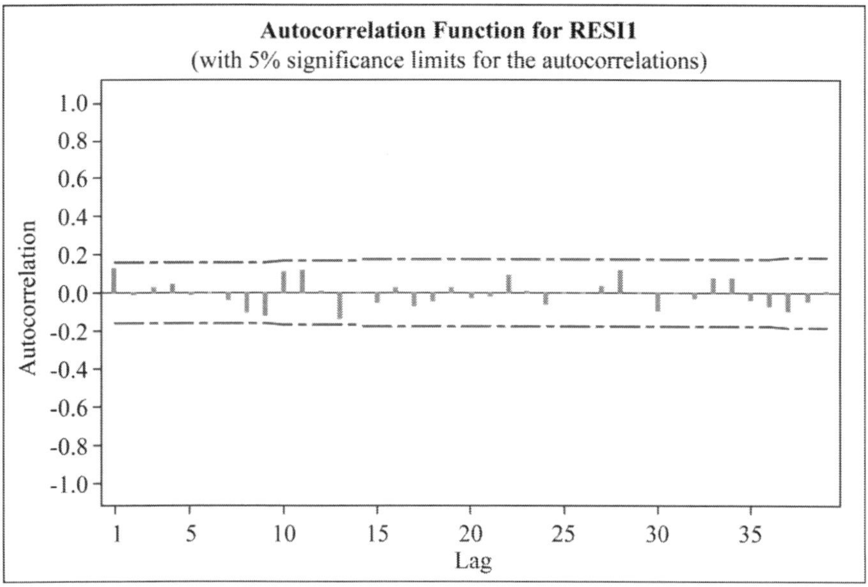

Figure 8. ACF of residuals.

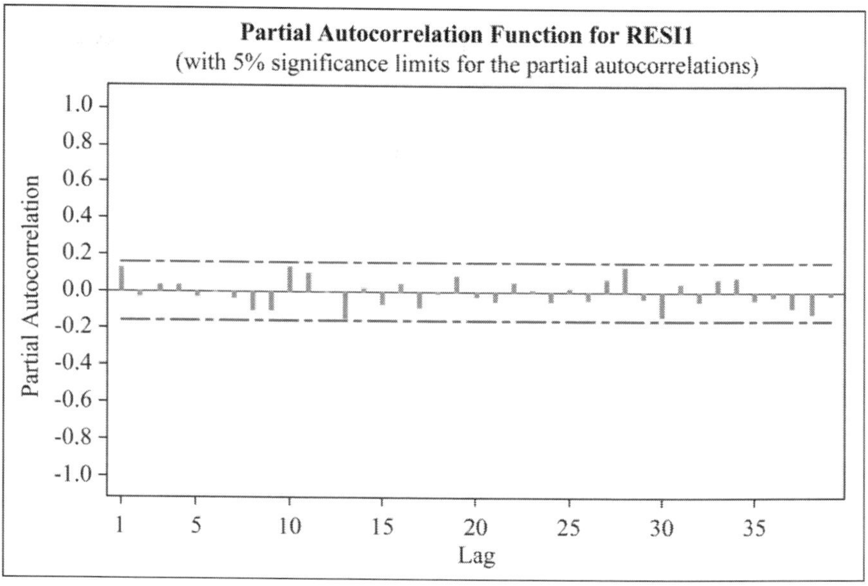

Figure 9. PACF of residuals.

A test of forecasting 10 months ahead in five distinct periods showed that, in all models, the error tends to increase after the second period, significantly impacting the average error. This fact could result from the choice of one model over another simply because of the errors from the second period are lower. This finding could be valid, if the prediction is used to making decisions in the midterm, in which the prediction of several periods ahead would determine actions to be taken and which could hardly be changed in the short term.

In the case of financial time series, significant changes in the forecasts made can trigger immediate decisions and corrections in a matter of minutes, or at worst a few days, repositioning strategies by hedging or even the rebalancing of the portfolio by the complete elimination of certain positions that would be affected or even entire investment strategies in a short space of time.

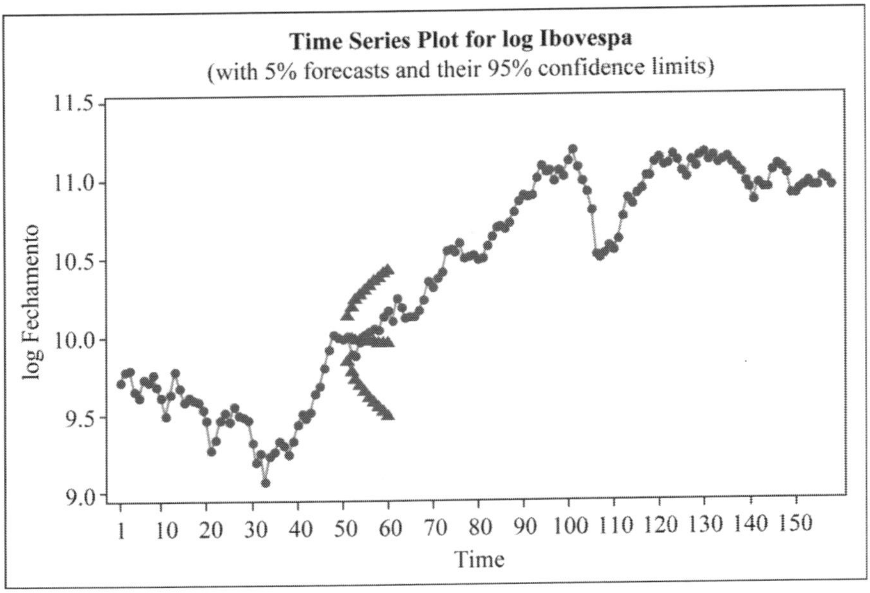

Figure 10. Forecast 10 points ahead, based on month 50.

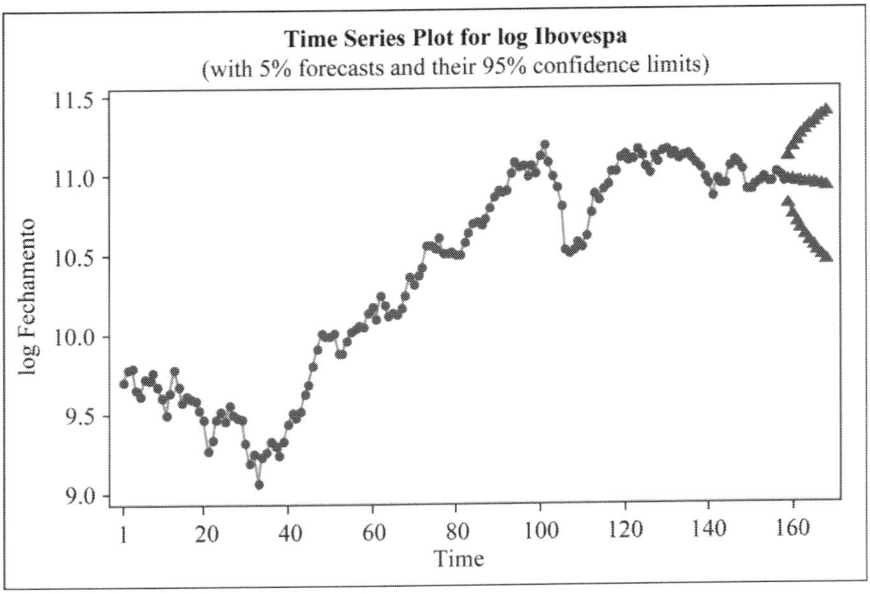

Figure 11. Forecast 10 points ahead, based on the last month.

Thus, it was chosen to perform the forecast one-step-ahead, in the case of this study, one month ahead. In the various models analyzed, this prediction proved to be such as that found the lowest MAPE. Thus, it avoids using only a period of the series, analyze the ASM one-step-ahead composed of five periods, which evaluated the forecast of each of these periods, being the MAPE constructed from the absolute percent average of the sum of these errors.

Table 3 shows the comparison of MAPE obtained by using four AR1 models, Single Exponential Smoothing, Double Exponential Smoothing and ARIMA (0, 2, 1). It is observed that the MAPE of the AR1 model enables its use when compared to other options.

5. CONCLUSIONS

Through the results obtained, it is observed that the model is effective in its forecasts. The statistics of the AR1 model coefficients and Chi-square statistics for modified Box-Pierce (Ljung-Box) provide proof to this fact.

A MAPE (mean absolute error percentage) of 0.052% was obtained, a lower value than those found in predictions made with other models used for comparison.

This study sought to obtain short-term forecasts for the next month (one step ahead) in order to minimize pre- diction errors. The model can be considered adequate for predicting the Bovespa Index series, and can be used as an aid to decision-making mechanism.

Table 3. MAPE table (absolute average percent error).

	AR1	Single Exp smoothing	Double Exp smoothing	ARIMA (0, 2, 1)
MAPE	0.052%	0.086%	0.118%	0.064%

ACKNOWLEDGEMENTS

Paulo Rotela Junior, Fernando Luiz Riêra Salomon, Edson de Oliveira Pamplona (National Counsel of Technological and Scientific Development), CAPES (Post-Graduate Federal Agency), and FAPEMIG (Foundation for the Promotion of Science of the State of Minas Gerais), which have been supporting the efforts for the development of this work in different ways and periods.

REFERENCES

1. Rotela, J.P., Pamplona, E.O. and Salomon, F.R. (2014) Otimização de Portfólios: Análise de Eficiência. RAE—Revista de Administração de Empresas, 54, 405-413.

2. Markowitz, H. (1952) Portfolio Selection. Journal of Finance, 7, 77-91.

3. Sharpe, W.F. (1963) A Simplified Model for Portfolio Analysis. Management Science, 9, 277-293.

4. Fama, E. (1970) Efficient Capital Markets: A Review of Theory and Empirical Work. Journal of Finance, 25, 383-417.

5. Slack, M., Chambers, S. and Johnston, R. (2007) Operations Management. Prentice Hall/Financial Times, Upper Saddle River.

6. Linstone, H.A. and Turoof, M. (1975) The Delphi Method: Techniques and Applications. Addison-Wesley, Boston.

7. Gaither, N. and Frazier, G. (2001) Operations Management. South-Western, Ohio.

8. Bacci, L.A. (2007) Combinação de métodos de séries temporais para a previsão da demanda de café no Brasil. Thesis, UNIFEI, Itajuba.

9. Morettin, P.A and Toloi, C.M. (2006) Análise de Séries Temporais—2ª Edição Revista e Ampliada. 2nd Edition, Editora Edgar Blucher, São Paulo.

10. Morettin, P.A. and Toloi, C.M. (1987) Previsão de séries temporais. 2nd Edition, Atual Editora, São Paulo.

11. Pindyck, R.S. and Rubinfeld, D.L. (2008) Microeconomics. 7th Edition, Prentice Hall, Upper Saddle River.

12. Makridakis, S., Wheelwrigt, S.C. and Hyndman, R.J. (1998) Forecasting: Methods and Applications. 3rd Edition, John Wiley & Sons, New York.

13. Fava, V.L. (2000) Metodologia de Box-Jenkins para Modelos Univariados. In: Vasconcellos, M.A.S. and Alves, D., Eds., Manual de econometria: nível intermediário, Atlas, São Paulo, 205-232.

14. Bertrand, J. and Fransoo, J. (2002) Operations Management Research Methodologies Using Quantitative Modeling. International Journal of Operations & Production Management, 22, 241-264.

15. Makridakis, S., Wheelwrigt, S.C. and Mcgee, V. (1983) Forecasting: Methods and Applications. 2nd Edition, John Wiley & Sons, New York.

CHAPTER 2

Time Series Modelling with Application to Tanzania Inflation Data

Edward Ngailo[1], Eliab Luvanda[2], Estomih S. Massawe[3*]

[1] Dares Salaam University College of Education, Dares Salaam, Tanzania

[2] Department of Economics, University of Dares Salaam, Dares Salaam, Tanzania

[3] Department of Mathematics, University of Dares Salaam, Dares Salaam, Tanzania

ABSTRACT

In this paper, time series modelling is examined with a special application to modelling inflation data in Tanzania. In particular the theory of univariate non linear time series analysis is explored and applied to the inflation data spanning from January 1997 to December 2010. Time series models namely, the autoregressive conditional heteroscedastic (ARCH) (with their extensions to the generalized autoregressive conditional heteroscedasticity ARCH (GARCH)) models are fitted to the data. The stages in the model building namely, identification, estimation and checking have been explored and applied to the data. The best fitting model is selected based on how well the model captures the stochastic variation in the data (goodness of fit). The goodness of fit is assessed through the Akaike Information Criteria (AIC), Bayesian Information Criteria (BIC) and minimum standard error (MSE). Based on minimum AIC and BIC values, the best fit GARCH models tend to be GARCH(1,1) and GARCH(1,2). After estimation of the parameters of selected models, a series of diagnostic and forecast accuracy test are performed. Having satisfied with all the model assumptions, GARCH(1,1) model is found to be the best model for forecasting. Based on the selected model, twelve months inflation rates of Tanzania are forecasted in sample period (that is from January 2010 to December 2010). From the results, it is observed that the forecasted series are close to the actual data series.

KEYWORDS

Time Series, Inflation, Autoregressive

1. INTRODUCTION

The concept of time series is based on the historical observations. It involves explaining past observations in order to try to predict those in the future [1]. A time series is a collection of observations measured sequentially through time. These measurements may be made continuously through time or be taken at a discrete set of time points [2].

Inflation as described by [3] is the persistent increase in the level of consumer prices or persistent decline in the purchasing power of money. Inflation can also be expressed as a situation where the demand for goods and services exceeds their supply in the economy [4]. In reality inflation means that your money can not buy as much as what it could have bought yesterday.

In recent years, inflation has become one of the major economic challenges facing most countries in the world especially those in Africa including Tanzania. Inflation is a major focus of economic policy worldwide as described by [5]. Inflation dynamics and evolution can be studied using a stochastic modelling approach that captures the time dependent structure embedded in the time series inflation data. The autoregressive conditional heteroscedasticity (ARCH) models, with its extension to generalized autoregressive conditional heteroscedasticity (GARCH) models as introduced by [6] and [7] respectively accommodate the dynamics of conditional heteroscedasticity (the changing variance nature of the data). Heteroscedasticity affects the accuracy of forecast confidence limits and thus has to be handled properly by constructing appropriate non-constant variance models [8].

The most common way of measuring inflation is the consumer price index (CPI) over monthly, quarterly or yearly. The inflation rate I_t at time t is calculated as

$$I_t = \frac{P_t - P_{t-1}}{P_{t-1}} \times 100\%$$

where

P_t is the current average price level of an economic basket of goods and services;

P_{t-1} is the average price level of the basket a year ago.

Within time series modelling, there are two approaches available for forecasting: the univariate and multivariate. In particular this paper will forecast future values of inflation time series data using the univariate forecasting approach, in which forecasts depend only on present and past values of a single series being forecasted.

2. CONDITIONAL HETEROSCEDASTICITY: ARCH-GARCH MODELS

Let $\{r_t\}$ be the mean-corrected return or rate of inflation, ε_t be the Gaussian white noise with zero mean and unit variance and H_t be the information set at time t given by $H_t = \{r_1, r_2, \ldots, r_{t-1}\}$. Then according to Engle [6], the process $\{r_t\}$ is ARCH(q) if

$$r_t = \sigma_t \varepsilon_t$$

where σ is the standard deviation and

$$E(r_t | H_t) = 0 \tag{1}$$

$$\text{Var}(r_t | H_t) = \sigma_t^2 = \alpha_0 + \sum_{i=1}^{q} \alpha_i r_{t-i}^2 \tag{2}$$

and the error term ε_t is such that

$$E(\varepsilon_t | H_t) = 0 \tag{3}$$

$$\text{Var}(\varepsilon_t | H_t) = 1 \tag{4}$$

with non-negativity condition that $\alpha_0 \geq 0$ and $\alpha_i \geq 0$ for all $i = 1, \ldots, q$.

The ARCH (1) model is a particular case of the general ARCH(q) model given by

$$\sigma_t^2 = \alpha_0 + \alpha_1 r_{t-1}^2, \tag{5}$$

with non-negativity condition that $\alpha_0 \geq 0$ and $\alpha_1 \geq 0$. α_0 and α_1 are unknown parameters.

2.1 Forecasting with the ARCH Model

The ARCH models provide good estimates of the series before it is realized. Let $r_1, r_2, r_3, \ldots, r_t$ be an observed time series. Then the l-step ahead forecast, for $l = 1, 2, \ldots$, at the origin t, denoted by $r_t(l)$, is taken to be the minimum mean squared error prediction, that is $r_t(l)$ minimizes $E(r_{t+l} - f(r))^2$, where $f(r)$ is a function of the observations. Then according to [9],

$$r_t(l) = E\left[r_{t+l} \middle| r_1, \ldots, r_t\right] \tag{6}$$

The forecasts for the r_t series do not provide helpful information. It is therefore more useful to consider the squared returns r_t^2 given by

$$r_t^2 = E\left[r_{t+l}^2 \middle| r_1^2, \ldots, r_t^2\right]$$

[10]. The l-step ahead forecast for the r_t^2 is given by

$$r_t^2(l) = E\left[r_{t+l}^2 \middle| r_t\right] = \widehat{\alpha_0}\left(1 + \widehat{\alpha_1} + \widehat{\alpha_1}^2 + \cdots + \widehat{\alpha_1}^{l-1}\right) + \widehat{\alpha_1}^l r_t^2 = \sigma_t^2(l). \tag{7}$$

The obvious possible problem in using the ARCH formulation is that the approach can lead to a highly parametric model if the lag q is large. This necessitates the use of the GARCH model as an extension to the ARCH model.

3. THE GARCH MODEL

A process $\{r_t\}$ is GARCH(p,q) if

$$r_t = \sigma_t \varepsilon_t,$$

$$\sigma_t^2 = \alpha_0 + \sum_{i=1}^{q} \alpha_i r_{t-i}^2 + \sum_{j=1}^{p} \beta_j \sigma_{t-j}^2 = \alpha_0 + \alpha(B) r_t^2 + \beta(B) \sigma_t^2 \tag{8}$$

where $\alpha(B)$ and $\beta(B)$ are polynomials in the backshift operator given by

$$\alpha(B) = \alpha_1 B + \cdots + \alpha_q B^q,$$

$$\beta(B) = \beta_1 B + \cdots + \beta_q B^q.$$

with the restrictions $\alpha_0 > 0$ $\alpha_i \geq 0$ and $\beta_j \geq 0$ for $i = 1,2,\cdots,q$ and $j = 1,2,\cdots,p$ being imposed in order to have the conditional variance remaining positive. Equation (8) can be expressed as

$$(1 - \beta(B))\sigma_t^2 = \alpha_0 + \alpha(B) r_t^2. \tag{9}$$

The GARCH(p,q) model does not show autocorrelation in the return series $\{r_t\}$. However the squared returns show autocorrelation even though the returns are not correlated. Writing r_t^2 in terms of $v_t = r_t^2 - \sigma_t^2$ yields

$$r_t^2 = \alpha_0 + \sum_{i=1}^{q} \alpha_i r_{t-i}^2 + \sum_{j=1}^{p} \beta_j r_{t-j}^2 - \sum_{j=1}^{p} \beta_j v_{t-j} + v_t. \tag{10}$$

Let $m = \max(p,q)$. Then

$$r_t^2 = \alpha_0 + \sum_{i=1}^{q} (\alpha_i + \beta_i) r_{t-j}^2 - \sum_{j=1}^{p} \beta_j v_{t-j} + v_t, \tag{11}$$

where $\alpha_i = 0$ for $i > q$ and $\beta_j = 0$ for $i > p$. Thus the equation of r_t^2 has an autoregressive moving average ARMA(m, p) representation.

In order to find the GARCH(p,q) process, we consider solving for α_0 in Equation (8) and let the variance of r_t be σ_ε^2, getting

$$\alpha_0 = \sigma_\varepsilon^2 \left(1 - \sum_{i=1}^{q} \alpha_i - \sum_{j=1}^{p} \beta_j\right). \tag{12}$$

Substituting Equation (12) into the Equation (11) one gets

$$r_t^2 = \sigma_\varepsilon^2 + \sum_{i=1}^{m} (\alpha_i + \beta_j)(r_{t-j}^2 - \sigma_\varepsilon^2) - \sum_{j=1}^{p} \beta_j v_{t-j} + v_t. \tag{13}$$

Therefore

$$r_{t-k}^2 - \sigma_\varepsilon^2 = \sum_{i=1}^{m} (\alpha_i + \beta_j)(r_{t-j}^2 - \sigma_\varepsilon^2) - \sum_{j=1}^{p} \beta_j v_{t-j} + v_t. \tag{14}$$

Multiplying both sides of Equation (14) by $(r_{t-k}^2 - \sigma_\varepsilon^2)$, and taking expectations we get

$$E\left[(r_{t-k}^2 - \sigma_\varepsilon^2)(r_t^2 - \sigma_\varepsilon^2)\right]$$
$$= E\left[\sum_{i=1}^{m} (\alpha_i + \beta_i)(r_{t-i}^2 - \sigma_\varepsilon^2)(r_{t-k}^2 - \sigma_\varepsilon^2)\right] - E\left[\sum_{j=1}^{p} \beta_j v_{t-j} (r_{t-k}^2 - \sigma_\varepsilon^2)\right] + E\left[v_t (r_{t-k}^2 - \sigma_\varepsilon^2)\right].$$

But

$$E\left[v_t (r_{t-k}^2 - \sigma_\varepsilon^2)\right] = E\left[(r_{t-k}^2 - \sigma_\varepsilon^2) E(v_t | r_t)\right] = 0,$$

and since v_t is a martingale difference, also

$$E\left[\beta_j v_{t-j} (r_{t-k}^2 - \sigma_\varepsilon^2)\right] = E\left[(r_{t-k}^2 - \sigma_\varepsilon^2) E(v_{t-j} | r_{t-k})\right] = 0 \text{ for } k < j.$$

Thus the autocovariance of the squared returns for the GARCH$^{(p,q)}$ model is given by

$$\operatorname{cov}\left(r_t^2, r_{t-k}^2\right) = E\left[\sum_{i=1}^{m}(\alpha_i + \beta_i)\left(r_{t-k}^2 - \sigma_\varepsilon^2\right)\left(r_{t-k}^2 - \sigma_\varepsilon^2\right)\right] = \sum_{i=1}^{m}(\alpha_i + \beta_i)\operatorname{cov}\left(r_t^2, r_{t-k+i}^2\right), \tag{15}$$

Dividing both sides of Equation (15) by $\operatorname{var}\left(r_t^2\right)$ gives the autocorrelation function at lag k as

$$\rho_k = \sum_{i=1}^{m}(\alpha_i + \beta_i)\rho_{k-i} \quad \text{for } k \geq p+1 \tag{16}$$

Letting ϕ_{mm} to denote the mth partial autocorrelation for r_t^2, Equation (16) can be written as

$$\rho_k = \sum_{i=1}^{m}\phi_{mi}\rho_{k-i}, \quad k = 1, \ldots, m \tag{17}$$

By Equation (17), ϕ_{mm} cuts off after lag q for an ARCH$^{(q)}$ process such that $\phi_{mm} \neq 0$ for $k \leq q$ and $\phi_{mm} = 0$ for $k > q$. This is identical to the partial ACF (PACF) for an AR$^{(q)}$ process and decays exponentially [11].

Assuming $r_1, \ldots, r_q$ and $\sigma_1^2, \ldots, \sigma_p^2$ are known, the conditional maximum likelihood estimates of the GARCH Model can be obtained by maximizing the conditional log-likelihood given by

$$\begin{aligned} l &= \log f\left(r_{q+1}, \ldots, r_t, \sigma^2_{p+1}, \ldots, \sigma^2_t \,\middle|\, \theta, r_1, \ldots, r_q, \sigma^2_1, \ldots, \sigma^2_p\right) \\ &= -\frac{1}{2}\sum_{t=m+1}^{T}\log\left(2\pi\sigma_t^2\right) - \frac{1}{2}\sum_{t=m+1}^{T}\left\{\frac{r_t^2}{\sigma_t^2}\right\}. \end{aligned}$$

with $\theta = \left(\alpha_0, \alpha_1, \ldots, \alpha_p; \beta_1, \beta_2, \ldots, \beta_q\right)$, and $m = \max(p, q)$.

3.1 Forecasting with GARCH(p,q) Models

The l-step ahead forecast of the conditional variance in a GARCH$^{(p,q)}$ model is given by

$$\sigma_t^2(l) = E\left[r_{t+1}^2 \middle| r_t\right] = \alpha_0 + \sum_{i=1}^{m}(\alpha_i + \beta_i)E\left(r_{t+l-i}^2 \middle| r_t\right) - \sum_{i=1}^{p}\beta_i\left(v_{t+l-i} \middle| r_t\right). \tag{18}$$

where $E\left(r_{t+l-i}^2 \middle| r_t\right)$ for $i<l$ is given recursively by

$$\begin{cases} E\left(v_{t+l-i} \middle| r_t\right) = 0 & \text{for } i<l \\ E\left(v_{t+l-i} \middle| r_t\right) = v_{t+l-i} & \text{for } i \ge l \end{cases}$$

4. DATA ANALYSIS

This section is dedicated to fitting the GARCH family of models to the Tanzania inflation rate data which we obtained from the Tanzania National Bureau of Statistics. The original data set consist of 168 monthly observations of the Tanzania inflation rates spanning from January 1997 to December 2010 as shown in Table 1.

4.1. Pre-Estimation Analysis

To avoid the difficulties of possible premature convergence we perform a pre-fit analysis. This will lead to selecting the appropriate model that adequately describes the data. In this pre-estimation or pre-fit analysis, data are loaded in the form of a price series, and then converted to a return series (stabilized series). The pre-fit analysis checks the return series for correlation and then quantifies the correlation. Because GARCH modelling assumes a return series, we need to convert inflation data (raw data) to returns. Figure 1 below displays raw data of inflation rate and Figure 2 displays the return series converted from the raw.

The returns appear to be quite stable over time and the transformations from Inflation rate data to returns has produced a stationary time series. The GARCH model assumes that return series is a stationary process. This may seem limiting, but the inflation data to return transformation is common and generally guarantees a stable data set for GARCH modelling.

According to [6] any autocorrelations in the series have to be removed before a GARCH model is constructed. This is done by regressing the squares of the series r_t on its past squared values $r_t^2, r_{t-1}^2, \ldots$ with the number of lags determined by the form of the ACF and PACF. The figures below display the sample autocorrelation function (ACF) of the returns based on the assumption that all autocorrelations are zero beyond lag zero.

In Figure 3 and Figure 4, the ACF and PACF provide no indication of the correlation characteristics of the returns. The ACF of squared returns in Figure 5 shows significant correlation and die out slowly. These results indicate that the variance of returns series is conditional on its past history and may change over time.

Statistical test for heteroscedasticity is carried out in order to establish the presence of ARCH effects in the data. This is shown in Table 2, Table 3 and Table 4. This is done using Ljung-Box-Pierce Q-test and Engle's ARCH test ([6] [12]).

From Table 2 it can be seen that there is no significant correlation in the raw returns at the 5% level of significance since H = 0. However, there is significant serial correlation in the squared returns in Table 3 when tested with the same inputs.

Table 1. Summary for statistics for Tanzania's monthly inflation.

Period	Average	Standard Deviation
Jan 1997-Sept 2001	9.94	4.3
Oct 2001-July 2005	5.1	1.58
Aug 2005-April 2009	8.24	2.52
May 2009-Dec 2010	8.9	3.0
Overall Period		
Jan 1997-Dec 2010	8.1	3.7

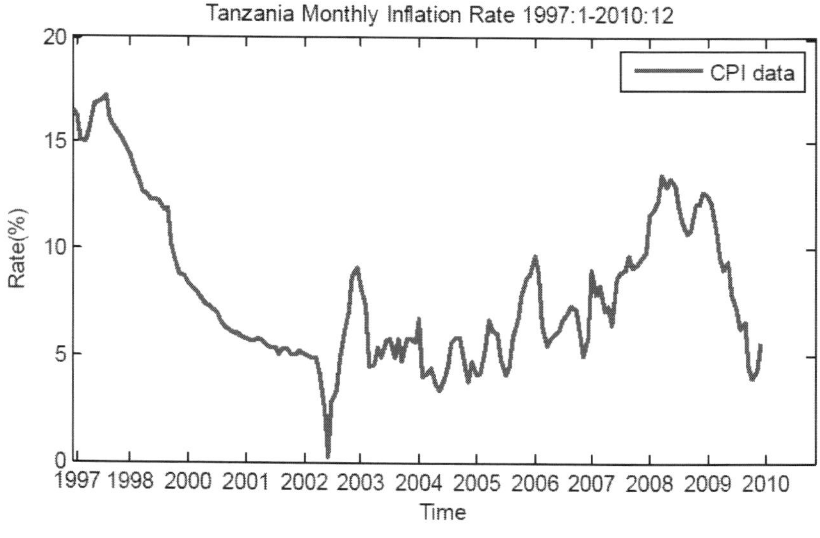

Figure 1. Time plot of monthly inflation in Tanzania.

Table 4 is the Engle's test for return series which shows that there is a significant correlation in the series, indicating presence of ARCH effects that is heteroscedasticity. Each of these tests extracts the sample mean from the actual inflation series.

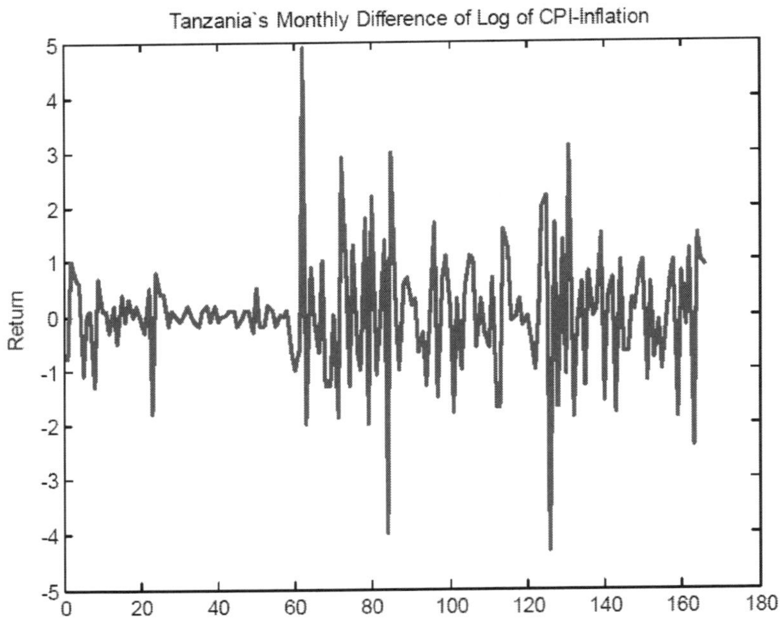

Figure 2. First difference of Log of CPI.

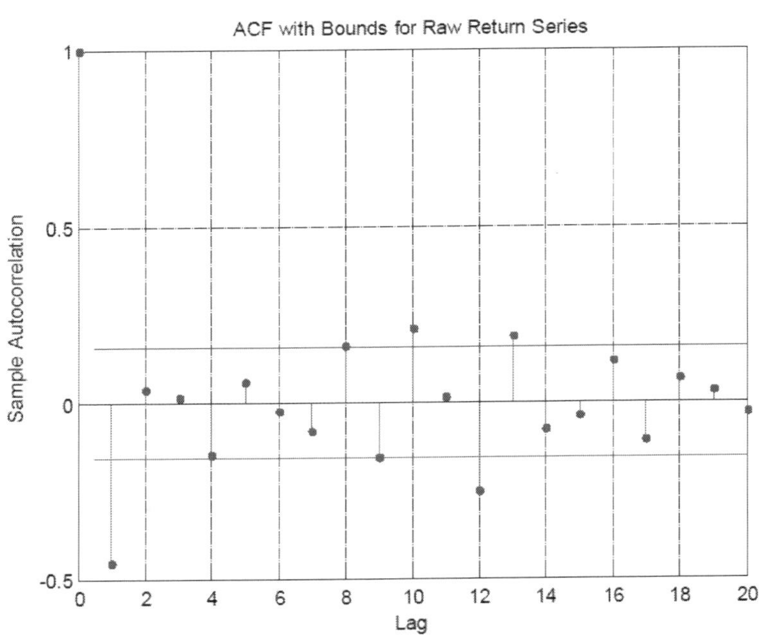

Figure 3. ACF with bounds for raw return series.

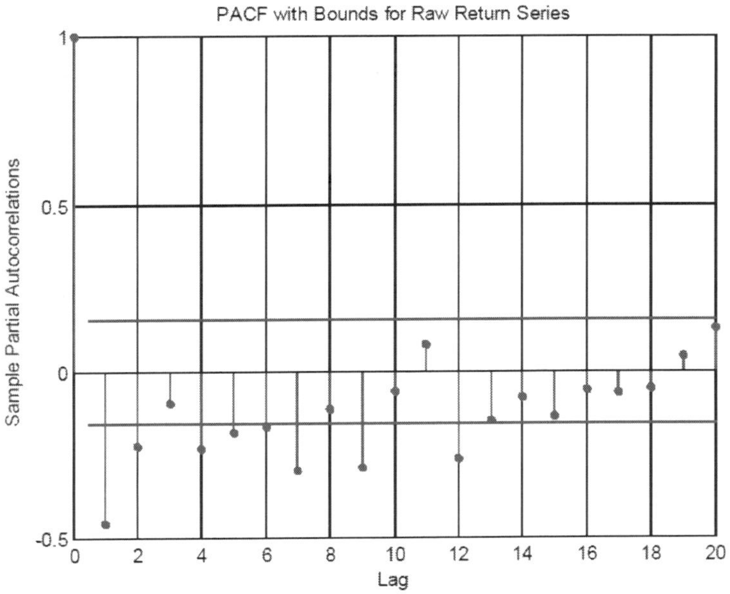

Figure 4. PACF with bounds for the raw returns series.

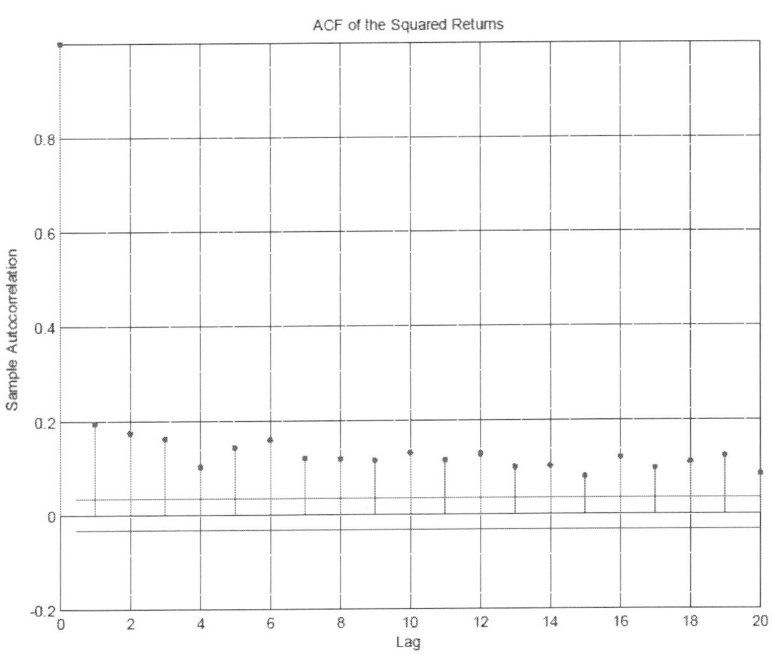

Figure 5. ACF of the squared returns.

Table 2. Ljung-Box-Pierce Q-test for autocorrelation (at 95% confidence).

Lag	H	p-value	Stat	Critical Value
10	0	0.0721	17.1019	18.3070
15	0	0.2359	18.529	24.9958
20	0	0.3888	21.1416	31.4104

Table 3. Ljung-Box-Pierce Q-test for squared returns (at 95% confidence).

Lag	H	P-value	Stat	Critical Value
10	1.000	0.0000	57.3782	18.3070
15	1.000	0.0000	76.7057	24.9958
20	1.000	0.0000	82.7525	31.4104

Table 4. Engle ARCH test for heteroscedasticity (at 95% confidence).

Lag	H	p-value	Stat	Critical value
10	1.000	0.0000	68.6467	18.3070
15	1.000	0.0000	67.9727	24.9958
20	1.000	0.0000	66.2913	31.4104

4.2. Model Estimation and Evaluation

4.2.1. Model Selection

The strategy used in selecting the appropriate model from competing models is based on the Akaike information criterion (AIC), the Bayesian information criterion (BIC) and standard error (MSE) and on the significance tests.

MATLAB software is used to perform trial and error evaluations to determine the best fitting model. The idea is to have a parsimonious model that captures as much variation in the data as possible. Usually the simple GARCH model captures most of the variability in most stabilized series. Small lags for p and q are common in applications. Typically GARCH$^{(1,1)}$, GARCH$^{(2,1)}$ or GARCH$^{(1,2)}$ models are adequate for modelling volatilities even over long sample periods [7].

From the derived models, using the method of maximum likelihood the estimated parameters of GARCH$^{(1,1)}$ model is summarized in the Table 5:

$$r_t = 0.0272 + \varepsilon_t,$$

$$\sigma_t^2 = 0.0753 + 0.45730 r_{t-1}^2 + 0.54266 \sigma_{t-1}^2.$$

The standard errors are used to assess the accuracy of the estimates, the smaller the better. The model fit statistics used to assess how well the model fit the data are the AIC and BIC. The corresponding values are: AIC = 474.8 and BIC = 487.3 with the log likelihood function of 233.4. The standard errors are quiet small suggesting precise estimates. Based on 95% confidence level, the coefficients of the GARCH$^{(1,1)}$ model are significantly different from zero and the estimated values satisfy the stability condition, that is $\alpha + \beta < 1$.

Table 6 below summarizes fit statistics for the other GARCH models which were considered.

In conclusion, it can be established that, amongst all the identified models, the GARCH$^{(1,1)}$proves the best fit model.

4.2.2. Diagnostic Checking of the GARCH(1,1) Model

One of the assumptions of GARCH models is that, for a good model, the residuals must follow a white noise process. Figure 6 inspects the relationship between the innovations (residuals) derived from the fitted model, the corresponding conditional standard deviations and the observed returns.

It can be observed that both innovations and returns exhibit volatility clustering. However if we plot the, standardized innovations (the innovations divided by their conditional standard deviation), they appear generally stable with little clustering as seen in Figure 7.

The time plot of the residuals given in Figure 7 is used to check whether the residuals are random. The normality check is also done by analyzing the histogram of residuals and normal probability plot. Figure 8 gives the histogram of the residuals from the GARCH(1,1) model. The histogram shows almost a symmetric bellshaped distribution which is indicative of the residuals following a normal distribution. The slight negative skewness is expected since the residuals may also come from student's t distribution. The negative skewness tendency is also supported by negative large residuals in Figure 7.

Table 5. Parameter estimates for GARCH (1,1).

Parameter	C	K	GARCH (1)	ARCH(1)
Estimates	0.0272	0.0753	0.4573	0.5427
Standard Error	0.0283	0.0254	0.0681	0.0958
t-value	0.9627	2.9656	6.7136	5.6619
$(>\|t\|)$	0.8315	<0.0001	<0.0001	0.0001>

Table 6. Comparison of suggested GARCH models.

Model	AIC	BIC	MSE	Log-Likelihood
GARCH(0,1)	489.0200	498.3560	0.1026	241.5100
GARCH(1,1)	**474.8236**	**487.2715**	**0.0544**	**233.4118**
GARCH(0,2)	478.5589	491.0068	0.0860	235.2794
GARCH(1,2)	491.1419	491.1419	0.0863	232.7910
GARCH(2,1)	476.8236	492.3835	0.0735	233.4118
GARCH(2,2)	477.1047	495.7766	0.0654	232.5524

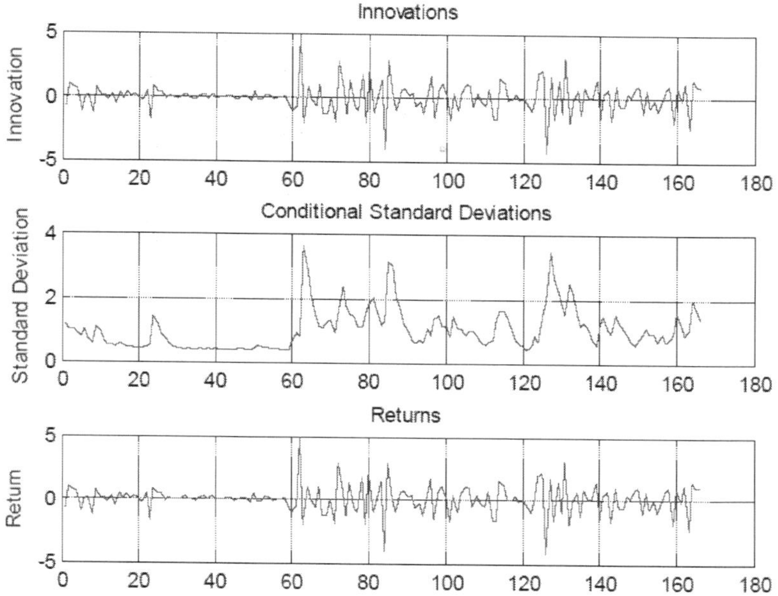

Figure 6. Plot for return, estimated volatility and innovations (residuals).

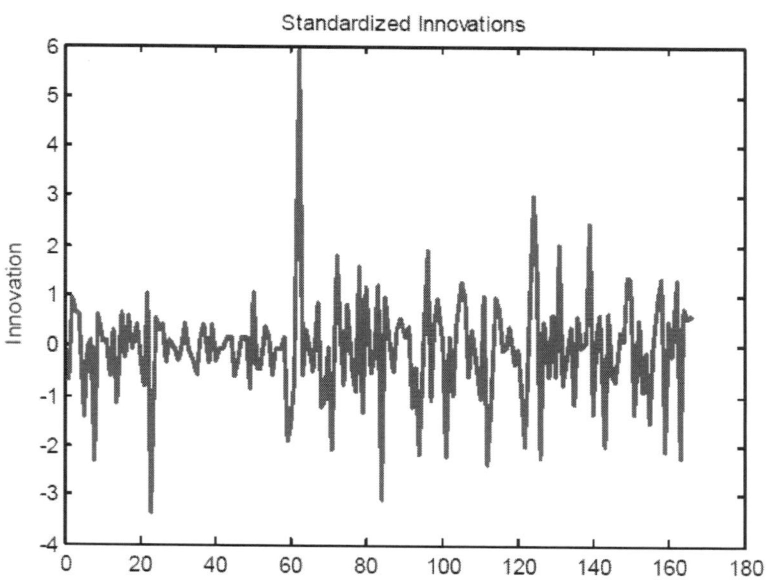

Figure 7. Time plot of residuals from GARCH(1,1).

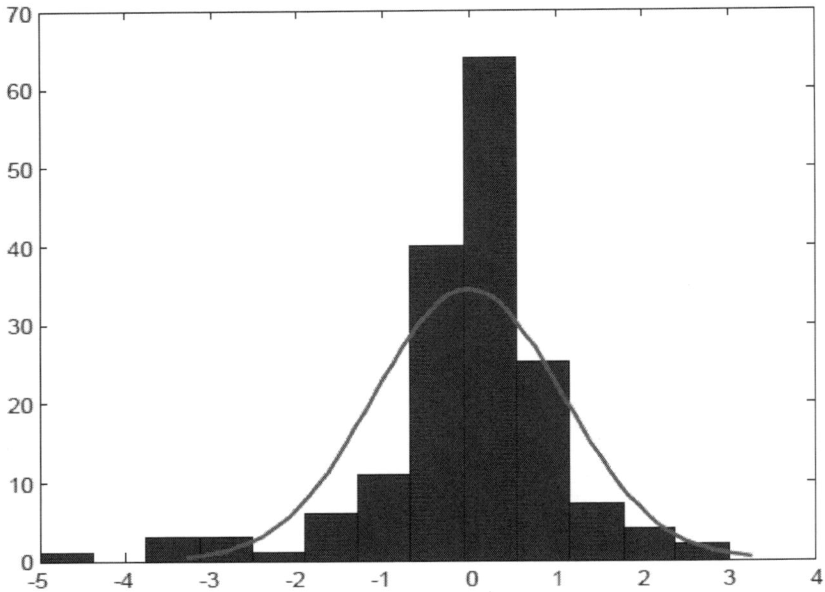

Figure 8. Histogram of residuals from GARCH (1, 1).

Figure 9 gives the plot of the ACF of the squared standardized innovations. The plot shows no correlation left. Table 7 and Table 8 compare the results of the Q-test and the ARCH test with the results of these same tests in the pre-estimation analysis in Table 2 and Table 3 respectively. In the pre-estimation analysis, both the Q-test and the ARCH test indicated rejection ($H = 1$with p value$= 0$) of their respective null hypothesis showing significant evidence in support of GARCH effects. In the post estimation analysis using standardized innovations based on the estimated model, these same tests indicate acceptance ($H = 0$with highly significant p-values) of their respective null hypothesis and confirm the explanatory power of GARCH$^{(1,1)}$. The tests showed that no any ARCH effects left (no heteroscedasticity).

4.3. Forecasting with the GARCH(1,1) Model

Table 7 shows the various measures of forecasting errors, namely the mean absolute error (MAE), the root mean squared error (RMSE), and Thiele's U test for two models seemed to be adequately suitable to fit the data. The first two forecast error statistics depend on the scale of the dependent variable. These are used as relative measure to compare forecasts for the same series across different models. The smaller the error, the better the forecasting ability of that model. The remaining two statistics are scale invariant. The Theil inequality coefficient always lies between zero and one, where zero indicates a perfect fit.

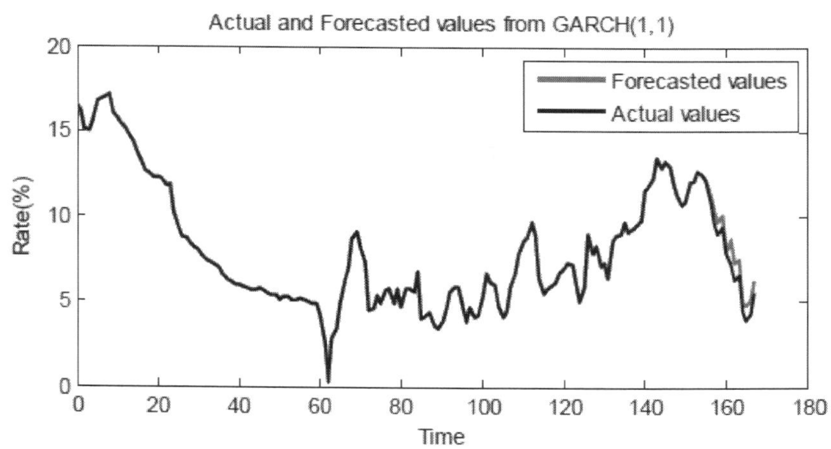

Figure 9. Time plot of inflation rate and one year forecasts by GARCH(1,1).

From Table 7, it can be seen that the accuracy test favour $\text{GARCH}^{(1,1)}$ model. Also the Thiele's statistics is less than one (0.8528) which indicates that, the forecasts are fairly accurate.

Table 8 below shows the forecast of Inflation by GARCH(1,1) for a period of one year from January 2010 to December 2010.

The Figure 9 displays the actual inflation rate and the predicted inflation rate by the GARCH$^{(1,1)}$model. The figure also displays how the forecasted values behave.

It can be observed from the Figure 9 that the forecasted inflation is closer to the actual inflation.

5. DISCUSSION AND CONCLUSION

In this paper, time series modelling was examined with a special application to modelling inflation data in Tanzania. In particular, the theory of univariate nonlinear time series analysis was explored and applied to the inflation data spanning from January 1997 to December 2010. The best fitting model was selected based on how well the model captures the stochastic variation in the data. Based on minimum Akaike Information Criteria (AIC) and Bayesian Information Criteria (BIC) values, it was observed that the best fit GARCH models were GARCH(1,1) and GARCH(1,2). However, after estimation of the parameters of selected models, a series of diagnostic and forecast accuracy test were performed and GARCH(1,1) model was found to be the best. Based on the selected model, twelve months inflation rates of Tanzania were forecasted in sample period (from January 2010 to December 2010). From the results, it is observed that the forecasted series are close to the actual data series.

Table 7. Forecast Accuracy Test on the most likely suggested GARCH models.

Model	MSE	MAE	RMSE	Thiele's U test
GARCH(1,1)	0.5848	0.7483	0.8651	0.8528
GARCH(1,2)	0.6034	0.9812	0.7768	0.9821

Table 8. Inflation forecast by GARCH(1,1) model for period of January to December 2010.

Month	Forecast (%)	Observed value (%)	Forecast error
January	11.32	10.9	0.42
February	10.43	9.6	0.83
March	9.66	9.0	0.66
April	10.08	9.4	0.66
May	7.90	7.2	0.70
June	8.74	7.9	0.84
July	7.29	6.3	0.99
August	7.48	6.6	0.88
September	5.03	4.5	0.53
October	4.81	3.9	0.91
November	5.14	4.3	0.84
December	6.30	5.6	0.70

REFERENCES

1. Ahiati, V.S. (2007) Discrete Time Series Analysis with ARMA Models. Revised Edition, Holden-Day, Oakland.

2. Chatfield, C. (2000) Time Series Forecasting. Chapman and Hall, London.

3. Webster, D. (2000) Webster's New Universal Unabridged Dictionary. Barnes and Noble, New York.

4. Hall, R. (1982) Inflation, Causes and Effects. Chicago University Press, Chicago.

5. David, F.H. (2001) Modelling UK Inflation, 1875-1991. Journal of Applied Econometrics, 16, 255-275.

6. Engle, R. (1982) The Use of ARCH/GARCH Models in Applied Econometrics. Journal of Economic Perspectives, 15, 157-168.

7. Bollerslev, T. (1986) Generalized Autoregressive Conditional Heteroscedasticity. Journal of Econometrics, 31, 307- 327.

8. Amos, C. (2010) Time Series Modelling with Application to South African Inflation Data. Master's Thesis, University of Kwazulu Natal, Kwazulu Natal.

9. Tsay, R.S. (2002) Analysis of Financial Time Series. John Wiley & Sons, Hoboken.

10. Shephard, N. (1996) Statistical Aspects of ARCH and Stochastic Volatility Time Series in Econometrics, Finance and Other Fields. Chapman and Hall, London.

11. Bollerslev, T., Chou, R.Y. and Kroner, K.F. (1992) ARCH Modeling in Finance: A Selective Review of the Theory and Empirical Evidence. Journal of Econometrics, 52, 5-59.

12. Ljung, G.M. and Box, G.E.P. (1978) On a Measure of Lack of Fit in Time Series Models. Biometrika, 65, 297-303.

CHAPTER 3

Financial Time Series Modelling of Trends and Patterns in the Energy Markets

Jane Aduda[1], Patrick Weke[2], Philip Ngare[2], Joseph Mwaniki[2]

[1] Department of Statistics and Actuarial Sciences, Jomo Kenyatta University of Agriculture and Technology, Nairobi, Kenya

[2] School of Mathematics, University of Nairobi, Nairobi, Kenya

ABSTRACT

Precise recognition of a time series path is important to policy makers, statisticians, economists, traders, hedgers and speculators alike. The correct time series path is also a key ingredient in pricing models. This study uses daily futures prices of crude oil and other distillate fuels. This paper considers the statistical properties of energy futures and spot prices and investigates the trends that underlie the price dynamics in order to gain further insights into possible nuances of price discovery and energy market dynamics. The family of ARMA-GARCH models was explored. The trends depict time varying variability and persistence of oil price shocks. The return series conform to a constant mean model with GARCH variance.

KEYWORDS

Financial Time Series, Trends and Patterns in Energy Markets, Futures and Spot Prices, ARCH Effects, ARMA-GARCH Models

1. INTRODUCTION

Energy use is behind virtually everything a person comes into contact with. The energy industry has rapidly expanded and become increasingly interdependent. In developed economies, the increase in energy consumption indicates a reliance on energy and its related products for continued and sustainable economic growth and development. Developing economies also rely on the development of energy resources to drive their growth. Energy was once viewed just as a utility, and an enabler with limited consumer interest, but now, it is key in the struggle for sustainable future economic growth [1] [2].

Energy prices, which are largely linked to oil prices, are a major concern for most economies. The recent financial crises and their ripple effects and after shocks have been largely unprecedented in terms of timing, speed and magnitude of impact on the world economies. Forecasting of crude oil prices is important for better investment and risk management and policy development, and econometric models are the most commonly used. Various authors have employed time series [3] [4], financial [5] and structural models [6] in forecasting crude oil prices. Financial studies involve asset returns [7] - [9], instead of prices, since, firstly, returns give a complete and scale-free summary of the investment opportunities and secondly, return series have nice statistical pro- perties known as stylised facts. These properties have important implications in model selection. Some of these interesting phenomena include volatility clusters which relate to the observation that the magnitudes of volatili- ties of financial returns tend to cluster together [9] [10], the leverage effect which relates to the asymmetry of news impact [11], leptokurticity, calendar effects and non-linear dependence.

In this paper, we regard an observed price series, $p_1, p_2, \cdots, p_T$, as a particular realisation of a stochastic process $\{p_t\}$. Standard time series analysis rests on important concepts such as stationarity, ergodicity, auto-correlation, white noise, innovation, and on a central family of models, the autoregressive moving average (ARMA) models [12] [13]. A stationary time series can be usefully described by it's mean, variance and autocorrelation or spectral density function [12] [14], and the joint probability distribution of any two observations, say, p_t, and p_{t+k} is the same for any two time periods t and $t+k$ separated by the same lag k [15]. For an ergodic process, we need not observe separate independent replications of the entire process in order to estimate its mean value or other moments [16].

In addition to the ARMA models, for the analysis of financial time series, we introduce the concept of volatility, which is pivotal in finance. Along side volatility we also consider the main stylised facts concerning financial time series [8] [13] [17]. In order to capture these stylised facts, we fit the autoregressive conditional heteroscedastic (ARCH) model first developed by Engle [18] and later generalized by Bollerslev [19] to generalised autoregressive conditional heteroscedastic (GARCH) model. A GARCH model is a parametric error distribution, which does not only capture volatility clustering but also accommodates some of the thick tails commonly found in financial time series.

This paper details the empirical analysis of financial time series data. We consider the official daily prices from the trading floor of the New York Mercantile Exchange (NYMEX) for a specific delivery month for Cushing Oklahoma West Texas Intermediate (OK WTI), Reformulated Blendstock for Oxygenate Blending (RBOB), and the number 1 heating oil futures contracts at close of business at 2:30 p.m. We begin the analysis from model identification, selection, estimation, diagnostic checking and finally forecasting. The analysis is done using MATLAB [20]. The rest of the paper is organized as follows: Section 2 discuses the methods used in empirical analysis, Section 3 gives a brief discussion of the data used, and the empirical analysis and the results obtained and finally Section 4 gives a summary of the work and the findings.

2. METHODOLOGY

2.1. Models

The full model under consideration falls under the general class of the combined $ARMA(p,q)-GARCH(P,Q)$ models which include an ARMA specification in the conditional mean and a GARCH specification in the conditional variance. The application of $ARMA(p,q)-GARCH(P,Q)$ model is a common approach in time series analysis that considers autocorrelation, volatility clustering, and heteroscedasticity. This class caters for models with constant variance under the $ARMA(p,q)$, and heteroscedastic models under the $GARCH(P,Q)$.

If $\{p_t\}$ denotes the price series at time t, let r_t be the log-return, and R_t, the relative or simple return so that

$$r_t = \log p_t - \log p_{t-1} = \ln\left(\frac{p_t}{p_{t-1}}\right) = \ln\left(1 + \frac{p_t - p_{t-1}}{p_{t-1}}\right) = \ln(1 + R_t), \quad (1)$$

an $ARMA(p,q) - GARCH(P,Q)$ model for the log-return series r_t, is specified as follows:

$$r_t = \sum_{i=1}^{p} \phi_i r_{t-i} + \sum_{i=1}^{q} \psi_i a_{t-i} + a_t \text{ and } a_t = \sigma_t \varepsilon_t \text{ where } \varepsilon_t \sim N(0,1)$$
$$\sigma_t^2 = \alpha_0 + \sum_{i=1}^{P} \alpha_i a_{t-i}^2 + \sum_{j=1}^{Q} \beta_j \sigma_{t-j}^2. \quad (2)$$

When $P = Q = 0$, the model reduces to an $ARMA(p,q)$. Other sub-models can be deduced depending on the respective values of the parameters p, q, P, Q.

2.2. Model Identification, Parameter Estimation and Model Selection

We start with an exploratory analysis of the price data to establish if the set of properties, common across many stocks, in many markets over time periods, that have been observed by many studies [7] [8] [17] are present. We examine spot and futures prices for these stylised facts and other statistical properties such as stationarity. We analyse the descriptive statistics and test for normality using the Jarque-Bera (JB) test [21].

To test for stationarity, the Dickey-Fuller (DF) test [22] is employed. If the data is non-stationary, dif- ferencing is performed until stationarity is achieved. Once stationarity is established, we study the autocorre- lation function (ACF) and partial autocorrelation function (PACF) in order to determine the form of the $ARMA(p,q)$ model for each series. A set of candidate models, around this initial model are then identified for further comparison.

Once the models for the conditional mean and variance have been identified from the differenced series, we employ the maximum likelihood estimation (MLE) method to fit parameters for the specified model of the dif- ferenced series.

The final model for each series is then selected based on Akaike's Information Criterion (AIC) which measures of goodness of fit or uncertainty for the range of values of the data. It measures the difference between a given model and the "true" underlying model. AIC is a function of the squared sum of errors (SSE), number of observations n, and the number of independent variables $k \le p+1$ where k includes the intercept, and is calculated as $AIC = n \cdot \ln\left(\frac{SSE}{n}\right) + 2k$. The best model is the one with the smallest AIC.

2.3. Forecasting

Model selection does not depend only on the goodness-of-fit of a model to the data, but also on the objective of the analysis. A model that is best in the in-sample fitting will not necessarily provide more accurate forecasts [8]. For this reason, it is common to use the performance of out-of-sample forecasts to aid in the selection of an adequate statistical model.

The forecasting ability of the model can only be determined by considering how well a model performs on data not used in estimating the model. It is common practice to partition the data into two sets, use the larger portion for estimating the model and the smaller one for testing the model. The test data can be used to measure model accuracy on new data. The size of the test data set should typically be about 20% of the total sample, although this depends on the sample size and the forecast horizon. In an ideal sense, the size of the test sample should be at least as large as the maximum forecast horizon required [23].

The sample standard deviation, σ of return is used as a simple forecast of volatility of returns, r_t, over the future period $[t, t+k]$. The k-days period historical variance is calculated as $\hat{\sigma}_t^2(k) = \frac{1}{k-1}\sum_{i=0}^{k-1}(r_{t-i} - \bar{r})$, where the sample mean return $\bar{r} = \frac{1}{k}\sum_{i=0}^{k-1} r_{t-i}$, is the estimate of the mean return μ. In most cases, the mean is usually set to zero in order to get a better forecast. Multiplying the variance by N, the number of trading days in a year and taking the square root gives the annualised volatility $\hat{\sigma} = \sqrt{N \cdot \hat{\sigma}^2}$. The number of trading days in a year is taken to be $N = 250$. $\hat{\sigma}$ is the best estimator for the volatility from the available price data and the volatility of any period of length, k, can be estimated from it.

The k-day volatility forecast can also be found by using the GARCH model. Under the condition that returns are uncorrelated across days, the k-day variance as of $t-1$ is given by

$$E\left[r_{t,t+k}^2 \mid \mathcal{F}_{t-1}\right] = E\left[r_t^2 \mid \mathcal{F}_{t-1}\right] + E\left[r_{t+1}^2 \mid \mathcal{F}_{t-1}\right] + \cdots + E\left[r_T^2 \mid \mathcal{F}_{t-1}\right] \quad (3)$$

The volatility forecast over the future period from $t+1$ to $t+k$, denoted by σ_{Ft}, is an average of the expected volatility on each day from t to $t+k$ i.e.

$$\sigma_{Ft} = \sqrt{\frac{1}{k}\sum_{i=1}^{k} E\left(\sigma_{t+k}^2 \mid \mathcal{F}_t\right)} \quad (4)$$

where

$$E\left(\sigma_{t+k}^2 \mid \mathcal{F}_t\right) = \alpha_0 \frac{1-(\alpha_1+\beta_1)^k}{1-(\alpha_1+\beta_1)} + (\alpha_1+\beta_1)^k \sigma_t^2. \quad (5)$$

Note that if $(\alpha_1 + \beta_1) < 1$,

$$\lim_{k\to 0} E\left(\sigma_{t+k}^2 \mid \mathcal{F}_t\right) = \frac{\alpha_0}{1-\alpha_1-\beta_1}, \tag{6}$$

which is the unconditional variance.

In-sample and out-of-sample forecasting ability of models can be measured using the mean absolute error (MAE) and root mean squared forecasting errors (RMSFE) which measures the out-of-sample losses. The RMSFE assigns greater weight to large forecast errors. This fact is handled using the MAE which on the contrary assigns equal weights to both over and under predictions of volatility.

The realised volatility at each forecast date is calculated from the expression $\sigma_{R,t} = \sqrt{\frac{1}{N}\sum_{i=1}^{N} r_{t+i}^2}$, so that, if the forecast volatility is σ_{Ft}, the RMSFE for a model is given by $RMSFE = \sqrt{\frac{1}{n}\sum_{t\in s}\left(\sigma_{Ft} - \sigma_{R,t}\right)^2}$ and the MAE is given by $MAE = \frac{1}{n}\sum_{t=1}^{n}\left|\sigma_{Ft} - \sigma_{R,t}\right|$ where n and s denote the number of forecasts and the set of times at which ex-ante forecasts are produced respectively in the above expression.

Forecasting using GARCH models is obtained recursively [24]. If we left t be the forecast origin, then, the k- step ahead forecast for σ_{t+k}^2 is

$$\sigma_t^2(k) = \hat{\alpha}_0 + \left(\hat{\alpha}_1 + \hat{\beta}_1\right)\sigma_t^2(k-1) \quad \text{for } k > 1. \tag{7}$$

An alternative way of writing Equation (7) is as shown in Equation (8)

$$\sigma_t^2(k) = \hat{\sigma}^2 + \left(\hat{\alpha}_1 + \hat{\beta}_1\right)^{k-1}\left(\sigma_t^2 - \hat{\sigma}^2\right) \quad \text{for } k > 1. \tag{8}$$

where

$$\hat{\sigma}^2 = \frac{\hat{\alpha}_0}{1 - \hat{\alpha}_1 - \hat{\beta}_1}.$$

From Equation (8), we can see that $\sigma_t^2 \to \hat{\sigma}^2$ as $k \to \infty$

3. DATA AND RESULTS

This paper investigates the trends and patterns of return series for spot and futures prices of Cushing OK WTI, RBOB and the number 1 heating oil traded in the NYMEX, for the period running from 2nd January 2006 to 22nd May 2015. The data was obtained from the US Energy Information Administration (EIA) website http://www.eia.gov/petroleum/data.cfm, accessed on June 25, 2015. The US EIA is the statistical and analytical agency within the US Department of Energy. It is the principal agency of the US Federal Statistical System responsible for collecting, analysing, and disseminating energy information to promote sound policy-making, efficient markets, and public understanding of energy and its interaction with the economy and the environment. EIA is the premier source of energy information in the US and, by law, its data, analyses, and forecasts are independent of approval by any other officer or employee of the US government. [25].

Figure 1 shows various series plots for daily spot and futures prices for Cushing OK WTI, RBOB gasoline and No. 1 heating oil traded at the NYMEX. The data appears non-stationary, with occasional jumps and spikes indicating heteroscedasticity. Testing the price series for stationarity using DF test shows that they all contain unit roots.

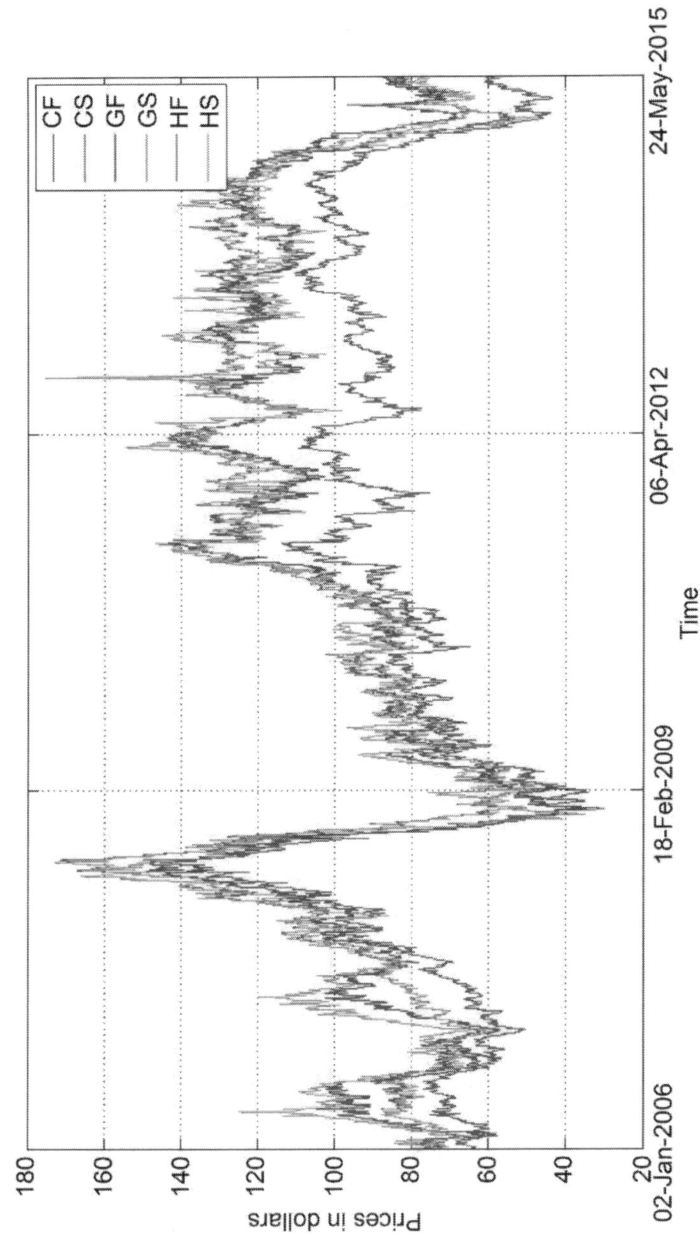

Figure 1. Time series plots for spot, futures prices and crack spreads in energy markets. CF = crude futures, CS = crude spot, GF = gasoline futures, GS = gasoline spot, HF = heating futures and HS = heating spot.

Table 1 gives a summary of the descriptive statistics and results for the JB test for the differenced price series. From the measure of skewness, all the series except the crude spot are skewed to the left. These series exhibit positive excess kurtosis, and these are some of the stylised facts observed in financial time series data. Based on the p-values of the JB test, we reject the null hypothesis of normality for all the differenced series.

Figure 2 shows plots of the differenced crude futures and spot prices. The series appear mean stationary but are not variance stationary. Volatility clustering, jumps and spikes are also observed.

We employ the DF test for stationarity on the differenced series. This is a one-sided left tail test andTable 2 gives the summary report of the DF test on the six series. Stationarity is confirmed for all these series, and hence they can be subjected to further time series analysis.

Figure 3 and Figure 4, show the ACF and PACF plots of crude futures and spot price series. From these plots, it is not easy to tell the form of $ARMA(p,q)$, significant lags are noticed up to lag 39 of the 40 plotted. The same behaviour is exhibited by the other four series. The ACF and PACF fail to identify the form of the model. We proceed to fit several $ARMA(p,q)$ models for $0 \leq p \leq 40$ and $0 \leq q \leq 40$, and the one with the smallest AIC is picked. Figure 5 shows a plot of AICs for various models fit for ΔCF, from which we see that the best model is an $ARMA(6,11)$ with an AIC of −4684.1. This model has the parameters estimated via the maximum likelihood, summarized in Table 3. For the crude spot data, the best model is an $ARMA(7,11)$. A similar analysis was also carried out on the differenced gasoline futures and prices and the results were similar with high orders of autoregressive and moving averages.

From Table 3, we can use the values of the t-statistic to check the significance of the parameters. The greater the absolute value of t, the greater the evidence against the null hypothesis that there is no significance dif- ference between the parameter estimate and 0. We can use the general rule of thumb that we reject the null hypothesis whenever $|t| \geq 2$.

After fitting the $ARMA(6,11)$ model, we standardized the inferred residuals and checked for normality.

Figure 6 shows two plots. The first one is the residual series which shows a lot of clustering, suggesting non- normality and non-constant variance. The second plot of Figure 6 is a histogram of the residuals also showing evidence of non-normality. Figure 7 shows a quantile-quantile plot and a box plot of the residuals and from these, we can also deduce non-normality of the residuals largely because of the extreme values.

The Kolmogorov-Smirnov (KS) test was employed to check normality of the residuals and the p-value was found to be <0.001 implying the residuals are not normally distributed. In this case the $ARMA(6,11)$ does not conform to the assumption normality in the innovations.

The Ljung-Box (LB) test on the residual series resulted in a p-value of 0.9781 indicating the residuals are not autocorrelated. Based on these results, we test for ARCH effects in the squared residuals using the LB test and obtained a p- value <0.001, indicating significant ARCH effects in the residuals.

Figure 8 also shows evidence of serial autocorrelation on the squared residuals. The PACF plot shows autoregression of order one so $ARCH(1)$ should capture the ARCH effects in the residuals.

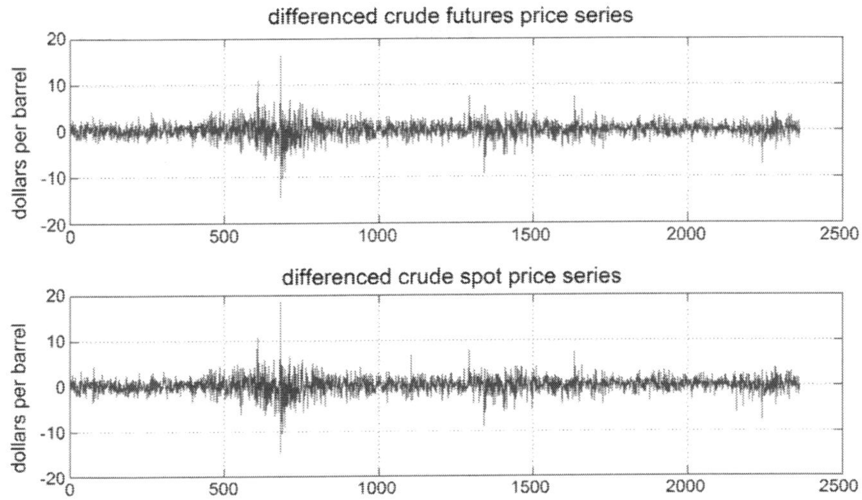

Figure 2. Differenced series for crude futures and spot prices.

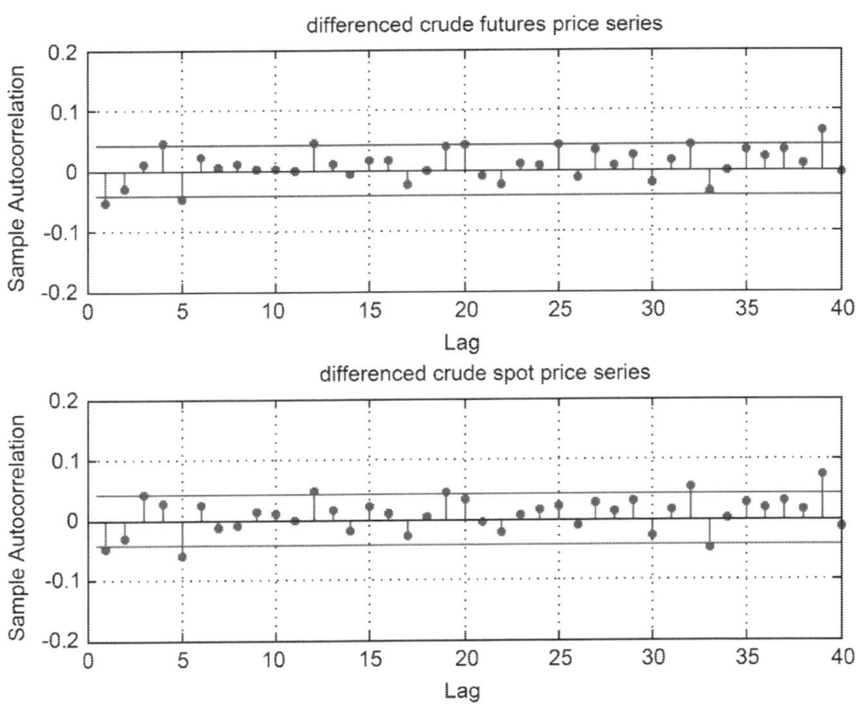

Figure 3. ACF plots for the differenced crude futures and spot price series.

Table 1. Descriptive statistics for energy spot and futures prices.

	ΔCF	ΔCS	ΔGF	ΔGS	ΔHF	ΔHS
Mean	−0.001	−0.002	0.005	0.003	0.003	0.001
Median	0.035	0.040	0.042	0.042	0.000	0.000
Maximum	16.370	18.560	13.020	21.336	12.306	11.508
Minimum	−14.310	−14.760	−17.724	−21.756	−17.304	−10.668
Skewness	−0.051	0.042	−0.456	−0.173	−0.467	−0.090
Kurtosis	10.252	12.046	8.008	8.312	8.699	5.837
JB *p*-value	0.001	0.001	0.001	0.001	0.001	0.001

Δ denotes differenced series.

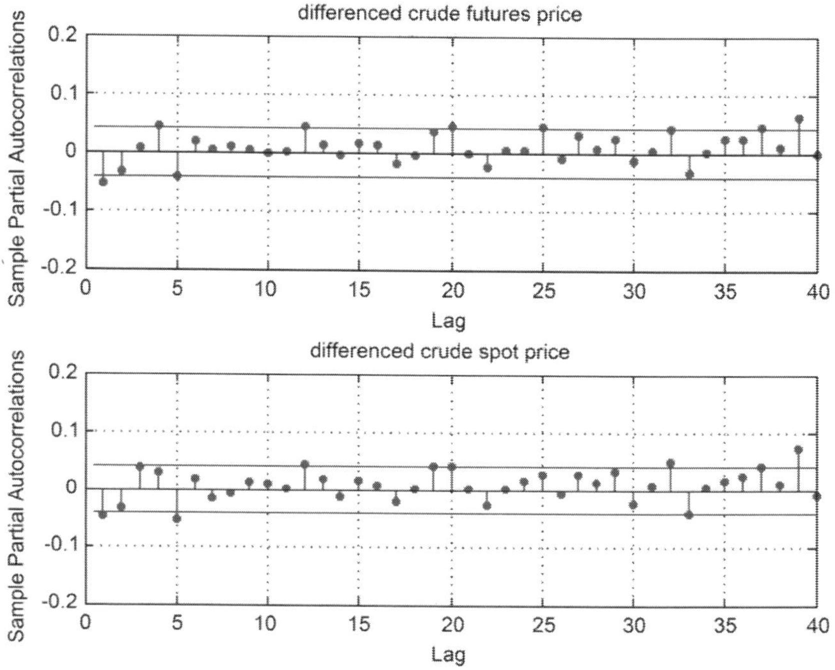

Figure 4. PACF plots for the differenced crude futures and spot price series.

We then specify a combined ARMA-GARCH model to capture the ARCH effects. Fitting a ARMA(6,11)- GARCH(1,1) had an AIC of 8810.9. However, a search for the best combined model revealed that an $ARMA(0,0)-GARCH(1,1)$ model gave the best fit with an AIC of −4.4037. From this analysis, we see that a more appropriate model to use would be one with a constant conditional mean and a conditional variance.

The analysis shows the differenced series exhibit "stylised facts" typically seen in high frequency financial data. volatility clustering manifests itself as autocorrelation in squared and absolute returns, or in the residuals of the estimated conditional mean equation. Examining the ACF and PACF plots for the squared differenced series shown in Figure 9 and Figure 10, reveal serial correlation of the squared differenced series. Gasoline and heating oil series exhibit the same behaviour.

The significance of these autocorrelations at various lags was tested using the LB test and the Lagrange multiplier (LM) test at lags 1, 5 and 10, and the results are summarized in Table 4. As is evident, there is enough evidence to reject the null hypothesis of no ARCH effects as per the LB test since the test indicates that there are significant ARCH effects in the all squared series. However, based on Engle's ARCH test, which is the LM test, we do not have sufficient evidence to reject the hypothesis of no ARCH effects for the gasoline futures and the heating oil futures series. For all the other series we reject that hypothesis of no ARCH effects.

Serial correlation in squared returns, or conditional heteroscedasticity can be modelled using GARCH models. GARCH models allow for the volatility to evolve with time. A GARCH model can be expressed as an ARMA model of squared residuals and hence many of it's properties follow easily from those of the corresponding ARMA process. If the fourth order moment of a GARCH (1,1) exists, the kurtosis implied by a GARCH (1,1) process with normal errors is greater than that of the normal distribution which is 3 [19]. For this reason, a GARCH model with normal errors can replicate some of the fat tailed behaviour observed in financial time series, though, most often, a GARCH model with a non-normal error distribution is required in order to fully capture the observed fat tail behaviour in financial time series.

We fit several $GARCH(p,q)$ model to the differenced crude futures series for $p = 20, q = 20$. We estimate the parameters α_0, α_i and β_j of the $GARCH(p,q)$ model and the in-sample estimates of the volatility. The Matlab command garch, models the return series as $GARCH(p,q)$ and estimates the parameters α_0, α_i and β_j via maximum likelihood.

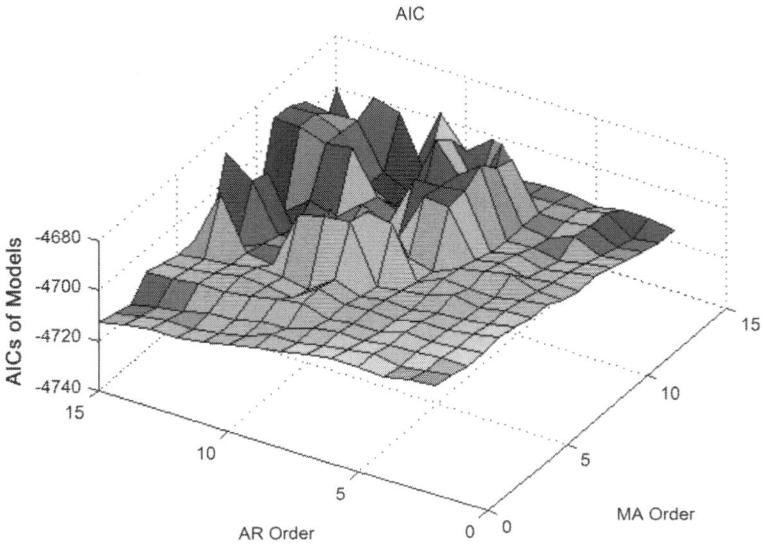

Figure 5. Plot for the AICs for the various ARMA models fitted for the differenced crude futures price data.

For the differenced crude futures series, the model with the smallest AIC was GARCH (19, 16) with an AIC of 8776.4 as shown in Figure 11 which shows the AICs of all the models fitted. The best model according to these fits have the parameters summarized in Table 5.

From these results, it can be seen that most of the in-between lags are not significant, the most important lags at 16 and 19. An analysis on the residuals of this model show that this model gives a fairly good fit. As per the AIC values, most these models do not seem to be significantly different and so the lags may be reduced to gain parsimony. An examination of the plots of the standardised residuals after fitting the GARCH(1,1) model for the return series indicates that the residuals are, on the overall, stable with some clustering. The standardised residuals exhibit no residual autocorrelation. It therefore makes more sense just to fit a $GARCH(1,1)$ for this data.

Table 2. Results from the Dickey-fuller test on the six series.

Series	*p*-value	Test statistic	Critical value
ΔCF	0.001	−51.2762	−1.9416
ΔCS	0.001	−50.9918	−1.9416
ΔGF	0.001	−49.5626	−1.9416
ΔGS	0.001	−50.9792	−1.9416
ΔHF	0.001	−50.0207	−1.9416
ΔHS	0.001	−50.6481	−1.9416

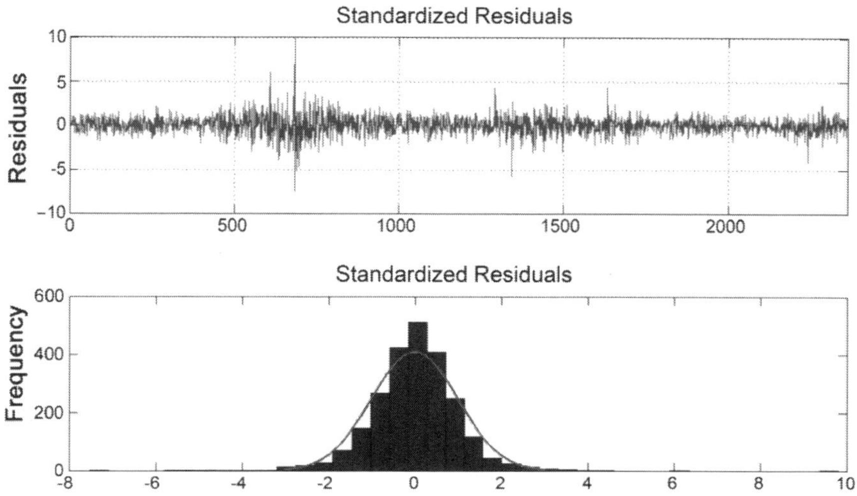

Figure 6. Normality plots for residual checks.

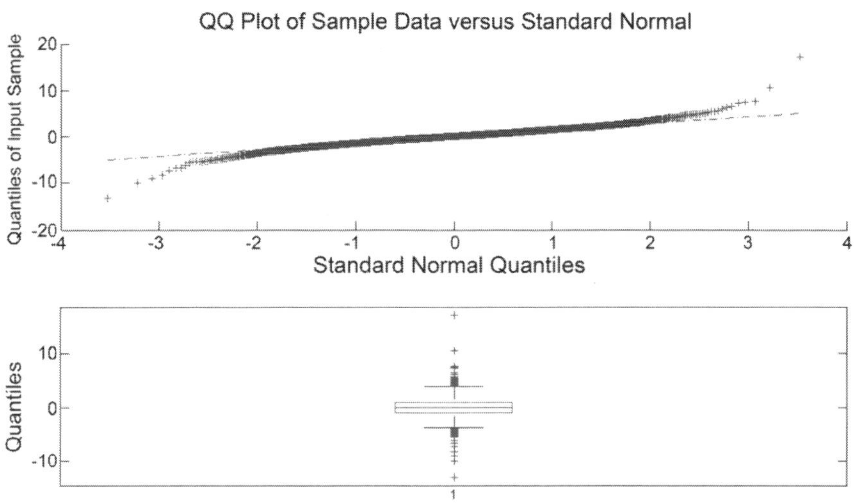

Figure 7. QQ-plot and Box plot for residuals from the ARMA(6, 11) model.

Table 3. Results from the ARMA(6, 11) Model fitted to the differenced crude futures price series.

Parameter	Estimate	Std Error	*t*-Statistic
Constant	−0.0003	0.0055	−0.0494
AR(1)	0.1064	0.2303	0.4622
AR(2)	0.5423	0.2234	2.4273
AR(3)	0.0277	0.0757	0.366
AR(4)	−0.7031	0.0729	−9.652
AR(5)	0.0806	0.2265	0.3558
AR(6)	0.8426	0.2163	3.896
MA(1)	−0.1625	0.2304	−0.7054
MA(2)	−0.5671	0.2374	−2.3886
MA(3)	0.0127	0.0852	0.1488
MA(4)	0.7652	0.0712	10.7484
MA(5)	−0.1797	0.2365	−0.7597
MA(6)	−0.8514	0.2529	−3.3666
MA(7)	0.0874	0.0380	2.2975
MA(8)	0.0383	0.0189	2.0253
MA(9)	−0.0402	0.0185	−2.1675
MA(10)	<0.001	0.0257	−0.001
MA(11)	0.0432	0.0242	1.7835
Variance	3.069	0.0503	61.0522

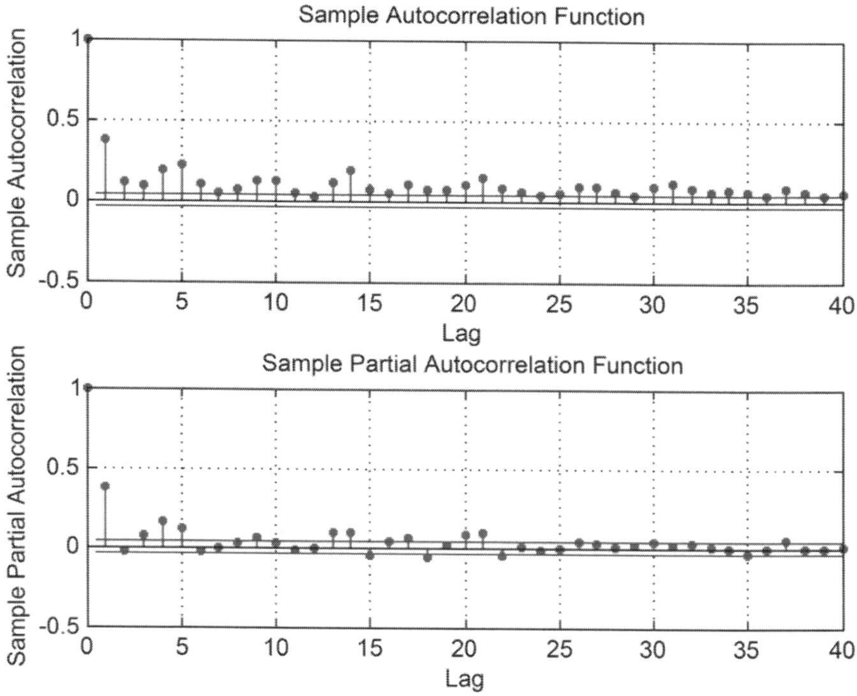

Figure 8. ACF and PACF for squared residuals from the ARMA(6,11) model.

The other series also behave in much the same way as the crude futures. Fitting a $GARCH(1,1)$ model on all the six series gives the results in Table 6. From the measure of persistence, we see the highest persistence of 0.9938 in the ΔCF series, although, all other return series are quite persistent. This indicates that the volatility processes return to their means after some time. The implication of this high persistence is that these shocks have a significant effect on the prices in these markets.

Forecasting

The total length of the data under consideration is 2363 data points. This data set was divided into two sets, the first sub-sample, used for training the model contained 2063 points and the second one, used for testing, con- tained 300 points.

In order to regain a forecast of the price series, the differencing on the forecast series obtained from the forecasts of the inferred conditional variances is undone. However because the inferred variances are con- ditioned on a standard normal error, there are several possibilities and, having many different forecasts and averaging would give a stable forecast for the crude futures prices, which can then be used to predict future crude oil futures prices. In Figure 12, we see one of the possibilities of the forecast prices. The RMSFE for this model is 0.2072, an indication that the forecasts are generally good.

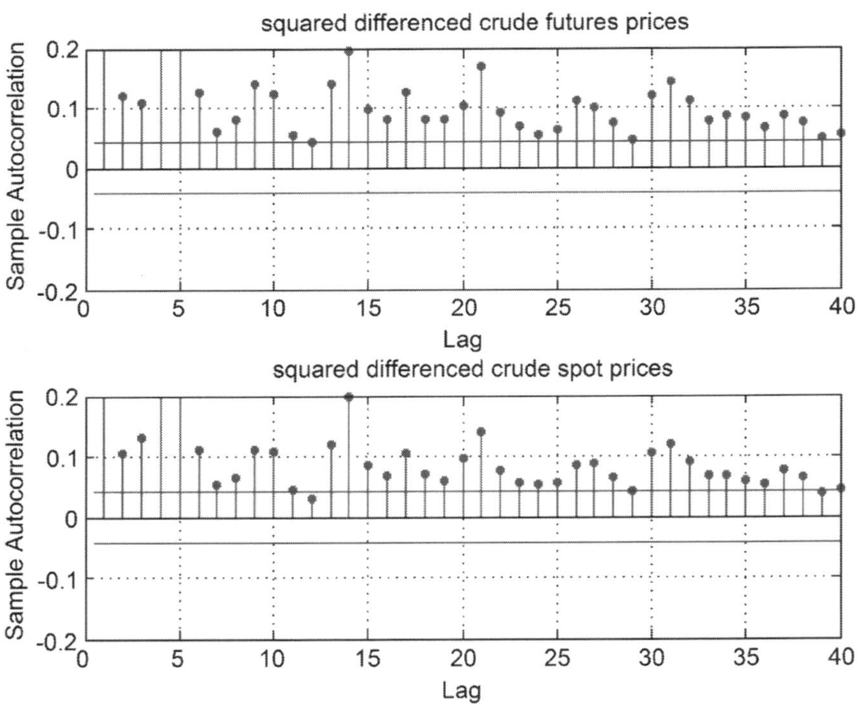

Figure 9. ACF plots for the squared differenced crude futures and spot price series.

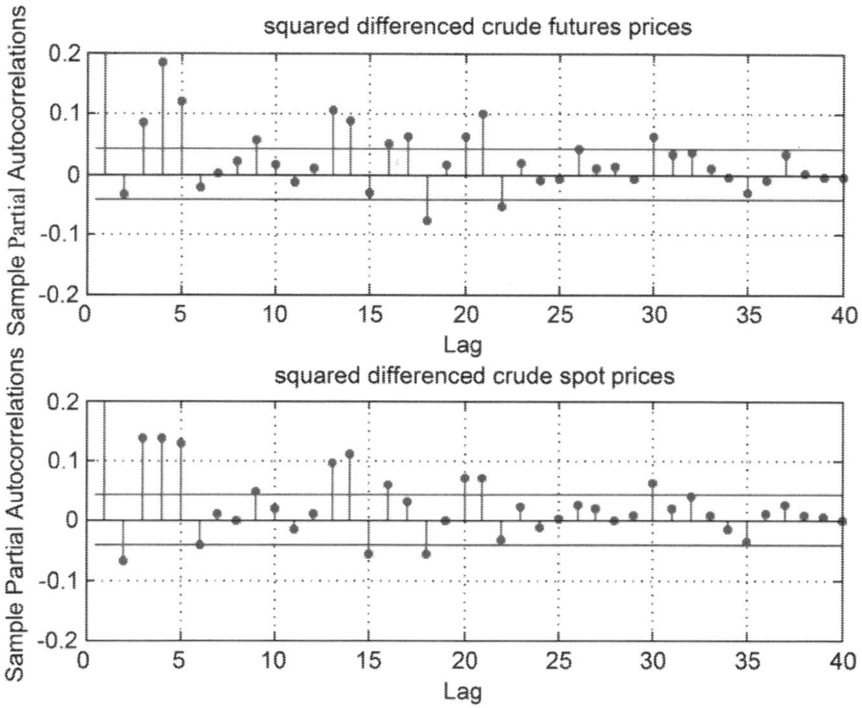

Figure 10. PACF plots for the squared differenced crude futures and spot price series.

4. DISCUSSION

This paper employs statistical and econometric techniques to investigate and model financial time series trends in energy markets. To do this, daily closing prices for a period of about 10 years for Cushing OK WTI, RBOB and number 1 heating oil spot and futures contracts traded in the NYMEX are considered. The paper also investigates the existence of stylised facts in these series, in order to fit an appropriate model that adequately describes the market dynamics.

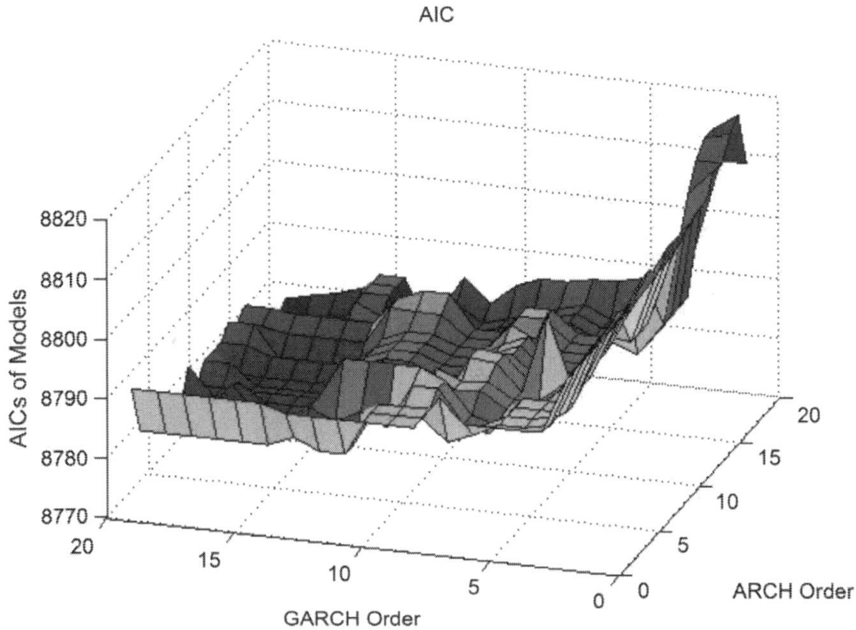

Figure 11. A plot for the AICs for various GARCH models fitted.

Price data are tested for stationarity using the DF test and results show the existence of unit roots. Normality is tested for using the JB test and non-normality is established. Return series are then generated from the price series through differencing, and then tested for stationarity using the DF test. The results reveal that return series are indeed mean stationary, but are definitely not variance stationary. Several ARMA models are fit to the return series and the standardised residuals analysed. For the crude futures return series, the best model turns out to be an $ARMA(6,11)$ an for the crude spot it is an $ARMA(7,11)$ among others, based on their AICs. These resultant models however contravene the Gaussian innovation assumption.

Table 4. Results from the Ljung-Box test and LM test for ARCH effects on the squared differenced series.

Lag	$Q(p)$			LM-test		
	1	5	10	1	5	10
$(\Delta CF)^2$	348.786	677.980	821.867	431.691	564.068	608.241
	(<0.001)	(<0.001)	(<0.001)	(<0.001)	(<0.001)	(<0.001)
$(\Delta CS)^2$	383.810	696.811	800.000	287.646	346.920	358.553
	(<0.001)	(<0.001)	(<0.001)	(<0.001)	(<0.001)	(<0.001)
$(\Delta GF)^2$	39.142	101.436	157.990	3.614	4.324	5.329
	(<0.001)	(<0.001)	(<0.001)	(0.0573)	(0.5038)	(0.8681)
$(\Delta GS)^2$	317.298	858.416	962.298	311.786	792.509	866.792
	(<0.001)	(<0.001)	(<0.001)	(<0.001)	(<0.001)	(<0.001)
$(\Delta HF)^2$	45.923	180.863	271.397	1.344	5.341	5.720
	(<0.001)	(<0.001)	(<0.001)	(0.2463)	(0.3758)	(0.8382)
$(\Delta HS)^2$	79.112	251.739	379.028	23.281	38.898	46.177
	(<0.001)	(<0.001)	(<0.001)	(<0.001)	(<0.001)	(<0.001)

p-values are indicated in the parenthesis.

Table 5. Results from the *GARCH*(19,16) model fitted to the differenced crude futures price series.

Parameter	Estimate	Std Error	*t*-Statistic
Constant	0.175	0.103	1.701
GARCH (16)	0.090	0.169	0.530
GARCH (19)	0.348	0.200	1.743
ARCH (1)	0.121	0.022	5.475
ARCH (2)	0.071	0.053	1.340
ARCH (5)	0.038	0.052	0.729
ARCH (6)	0.037	0.046	0.811
ARCH (8)	0.026	0.033	0.780
ARCH (9)	0.007	0.037	0.188
ARCH (10)	0.050	0.037	1.346
ARCH (13)	0.022	0.037	0.585
ARCH (14)	0.036	0.038	0.955
ARCH (15)	0.013	0.036	0.348
ARCH (16)	0.083	0.036	2.288

Table 6. Results from the $GARCH(1,1)$ model fitted to all the differenced price series.

Parameter	ΔCF	ΔCS	ΔGF	ΔGS	ΔHF	ΔHS
α_0	0.0191	0.0265	0.3354	0.5388	0.0743	0.0428
α_1	0.9464	0.9388	0.8236	0.8318	0.915	0.9429
β_1	0.0475	0.058	0.1072	0.1036	0.0642	0.0451
$(\alpha_1 + \beta_1)$	0.9938	0.9916	0.9308	0.9354	0.9792	0.9878

$(\alpha_1 + \beta_1)$ measures the persistence.

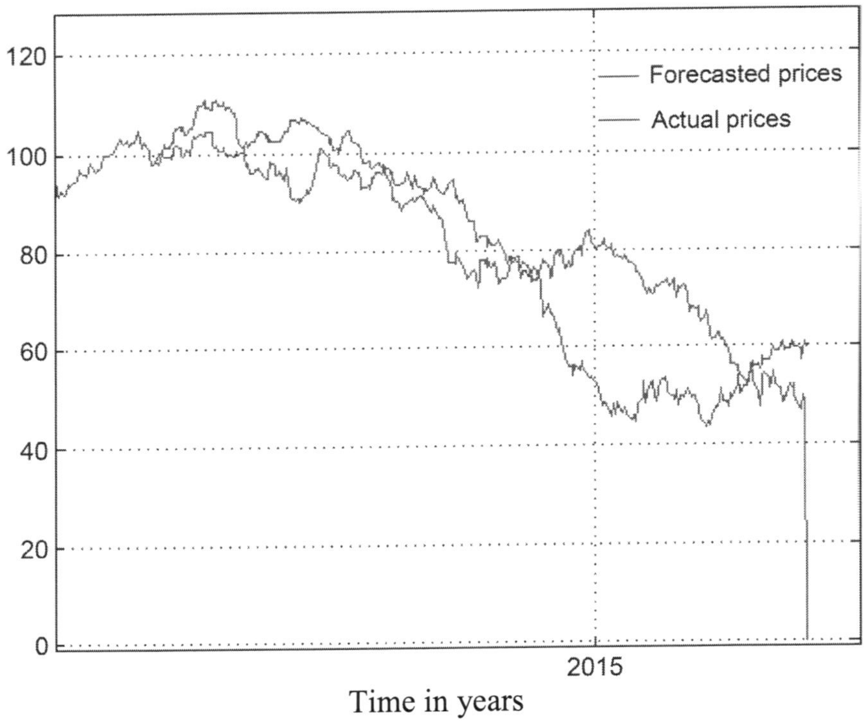

Figure 12. Crude futures price series showing foretasted prices.

We then propose a combined $ARMA(p,q)-GARCH(P,Q)$ model to capture the ARCH effects in the variance, then find that the best model under these circumstances is the $ARMA(0,0)-GARCH(1,1)$, implying a constant mean conditional variance equation. We finally find that for example, for the crude futures return series, the $GARCH(16,19)$ is the best model based on the AIC, although a $GARCH(1,1)$ does quite well for this and all the other six series with the residual analysis conforming to the assumption of Gaussian Innovations. GARCH models can therefore adequately model the trends and patterns in the energy markets. The trends also depict time varying variability and high persistence of oil price shocks. These shocks therefore have a significant impact on the prices of these energy prices.

REFERENCES

1. Dorsman, A., Simpson, J.L. and Westerman, W. (2013) Energy Economics and Financial Markets. Springer, Berlin.

2. Ramos, S.B. and Veiga, H. (2014) The Interrelationship between Financial and Energy Markets. Springer, Berlin.

3. Liu, H.P., Erdem, E. and Shi, J. (2011) Comprehensive Evaluation of ARMA-GARCH (-M) Approaches for Modeling the Mean and Volatility of Wind Speed. Applied Energy, 88, 724-732.

4. Beck, A., Shin, Y., Kim, A., Rachev, S., Feindt, M. and Fabozzi, F. (2013) Empirical Analysis of ARMA-GARCH Models in Market Risk Estimation on High-Frequency US Data. Studies in Nonlinear Dynamics and Econometrics, 17, 167-177.

5. Chinn, M.D., LeBlanc, M. and Coibion, O. (2005) The Predictive Content of Energy Futures: An Update on Petroleum, Natural Gas, Heating Oil and Gasoline. National Bureau of Economic Research, Cambridge.

6. Alvarez-Ramirez, J., Rodriguez, E., Martina, E. and Ibarra-Valdez, C. (2012) Cyclical Behavior of Crude Oil Markets and Economic Recessions in the Period 1986-2010. Technological Forecasting and Social Change, 79, 47-58.

7. Cont, R. (2001) Empirical Properties of Asset Returns: Stylized Facts and Statistical Issues. Quantitative Finance, 1, 223-236.

8. Tsay, R.S. (2010) Analysis of Financial Time Series. 3rd Edition, John Wiley & Sons, Inc., Hoboken.

9. Mandelbrot, B. (1963) The Variation of Certain Speculative Prices. The Journal of Business, 36, 394-419.

10. Engle, R.F. and Patton, A.J. (2001) What Good Is a Volatility Model. Quantitative Finance, 1, 237-245.

11. Black, F. (1976) The Pricing of Commodity Contracts. Journal of Financial Economics, 3, 167-179.

12. Box, G.E.P. and Jenkins, G.M. (1976) Time Series Analysis: Forecasting and Control, Revised Edition, Holden-Day.

13. Francq, C. and Zakoian, J.-M. (2011) GARCH Models: Structure, Statistical Inference and Financial Applications. John Wiley & Sons.

14. Hamilton, J.D. (1994) Time Series Analysis. Princeton University Press, Princeton.

15. Montgomery, D.C., Jennings, C.L. and Kulahci, M. (2011) Introduction to time Series Analysis and Forecasting. John Wiley & Sons, Hoboken.

16. Chan, N.H. (2011) Time Series: Applications to Finance with R and S-Plus. John Wiley & Sons, Hoboken.

17. Zivot, E. and Wang, J.H. (2007) Modeling Financial Time Series with S-PLUS. Springer Science & Business Media, Berlin.

18. Engle, R.F. (1982) Autoregressive Conditional Hetero-scedasticity with Estimates of the variance of United Kingdom Inflation. Econometrica, 50, 987-1007.

19. Bollerslev, T. (1986) Generalized Autoregressive Conditional Heteroscedasticity. Journal of Economics, 31, 307-327.

20. The MathWorks (2013) MATLAB Version 7.14.0.739 R2013a. The MathWorks Inc., Torrance.

21. Jarque, C.M. and Bera, A.K. (1987) A Test for Normality of Observations and Regression Residuals. International Statistical Review/Revue Internationale de Statistique, 55, 163-172.

22. Dickey, D.A. and Fuller, W.A. (1979) Distribution of the Estimators for Autoregressive Time Series with a Unit Root. Journal of the American Statistical Association, 74, 427-431.

23. Hyndman, R.J. and Koehler, A.B. (2006) Another Look at Measures of Forecast Accuracy. International Journal of Forecasting, 22, 679-688.

24. Jondeau, E., Poon, S.-H. and Rockinger, M. (2006) Financial Modeling under Non-Gaussian Distributions. Springer, Berlin.

25. EIA (2015) US Energy Information Administration.

CHAPTER 4

Forecasting High-Frequency Long Memory Series with Long Periods Using the SARFIMA Model

Handong Li, Xunyu Ye

School of Government, Beijing Normal University, Beijing, China

ABSTRACT

This paper evaluates the efficiency of the SARFIMA model at forecasting high-frequency long memory series with especially long periods. Three other models, the ARFIMA, ARMA and PAR models, are also included to compare their forecasting performances with that of the SARFIMA model. For the artificial SARFIMA series, if the correct parameters are used for estimating and forecasting, the model performs as well as the other three models. However, if the parameters obtained by the WHI estimation are used, the performance of the SARFIMA model falls far behind that of the other models. For the empirical intraday volume series, the SARFIMA model produces the worst performance of all of the models, and the ARFIMA model performs best. The ARMA and PAR models perform very well both for the artificial series and for the intraday volume series. This result indicates that short memory models are competent in forecasting periodic long memory series.

KEYWORDS

High-Frequency Financial Series, Long Memory, Long Periods, SARFIMA, Monte Carlo Simulation, Intraday Volume

1. INTRODUCTION

Recent years have witnessed a vast increase in the amount of high-frequency financial market data that are available. Using these data, practitioners are now able to manage their assets in greater detail. For example, the intraday volume series is often used in the Volume Weighted Average Price (VWAP) strategy to avoid a large reverse impact when executing large orders. Consequently, the econometrics of the high-frequency financial series receives wider attention in the academic field. As Engle (2005) summarizes, intraday financial series often contain periodic patterns and present a long horizon of strong dependence [1]. The autocorrelation function (ACF) of these series decays slowly and is particularly significant at the seasonal lags. These periods can be especially long when the sampling interval becomes short.

A number of works have been concerned with forecasting the periodic long memory series. They mainly focused on forecasting series with relatively short periods, namely twelve monthly periods or four seasonal periods in a year. On one hand, various long memory models have been used for forecasting this series. An autoregressive fractionally integrated moving average (ARFIMA) model [2] [3] was directly applied by Franses & Ooms (1997) in

[4] to forecast quarterly UK inflation. Porter-Hudak (1990) suggested a seasonal autoregressive fractionally integrated moving average (SARFIMA) model to forecast monetary aggregates [5]. This model tries to remove the hyperbolic decay at the seasonal lags by including a seasonal fractional differencing filter $(1-B^s)^D$ in the ARFIMA model, where B is the backward shift operator, s is the given period and D is the seasonal differencing parameters. This model is later used in [6] for monthly river flows and in [7] for inflation rates. By introducing seasonal dummy variables to seasonally change the fractional differencing parameter in the ARFIMA model, Ref. [4] proposed a periodic ARFIMA (PARFIMA) model for forecasting periodic long memory series. On the other hand, short memory models, such as the autoregression (AR) model and the periodic autoregression (PAR) model, were also proven to be competent in handling this series.

However, no consistent conclusion has been made on the superiority of specific models for forecasting periodic long memory series. Franses & Ooms (1997) [4] tried the periodic PAR model, AR model, PARFIMA model and ARFIMA model to forecast the quarterly UK inflation, but found no significant difference between these models. Those authors did find that the PARFIMA model was generally outperformed by rival models. Porter-Hudak (1990) compared the SARFIMA model and the Airline model, and found that the former outperformed the latter [5]. Nasr & Trabelsi (2005) tried the PARFIMA, SARFIMA, PAR, and AR models in [7] to forecast inflation rates in four different countries, and showed that the long memory models, the PARFIMA model and the SARFIMA model, performed better than the short memory models in terms of information criteria and clean residuals.

This paper studies the performance of different models when forecasting high-frequency long memory series with long periods. In particular, we want to deduce whether the SARFIMA model is capable of forecasting this type of series, because the mechanism of the SARFIMA process fits the description of the periodic long memory series well. Artificial SARFIMA series are generated to test the performance of different forecasting models, including the ARFIMA, the SARFIMA, the AR and the PAR models. We are also interested in finding a suitable model for forecasting intraday volume series, which is a very useful series for VWAP trading. These four models will also be tried on this series for comparison.

This paper is organized as follows: Section 2 introduces the four forecasting models that will be tested. Section 3 studies the performances of the four models through Monte Carlo simulations. Section 4 uses these models to forecast the intraday volume in both the American and Chinese stock markets and then compares their performance. Section 5 presents our conclusion.

2. THE MODELS

2.1. Long Memory Models

Two long memory models will be used in our study. The ARFIMA model ignores periodicity. The other, the SAFARMA model, includes periodicity.

If we assume that a simple fractional differencing operator can remove the high autocorrelation at both the seasonal lags and non-seasonal lags, we can use an ARFIMA model directly to help forecast a periodic long memory series. The ARFIMA model is defined as:

$$\phi(B)(1-B)^d (X_t - \mu) = \theta(B)\varepsilon_t$$

where μ is the mean of the process, ε_t is the white noise process, B is the backward shift operator, 0 < d < 0.5 is the differencing parameters respectively, $\phi(B)$, $\theta(B)$ are the polynomial operators with orders p, q respectively.

$$\phi(B)=1-\phi_1 B-\phi_2 B^2 - \text{L} - \phi_p B^p,\quad \theta(B)=1-\theta_1 B-\theta_2 B^2 - \text{L} - \theta_q B^q$$

A full definition of the SARFIMA $(p,d,q)\times(P,D,Q)_s$ model is defined as:

$$\phi(B)\Phi(B)(1-B)^d\left(1-B^s\right)^D(X_t-\mu)=\theta(B)\Theta(B)\varepsilon_t$$

where $s\in \mathrm{N}$ is the seasonal period, $0<d<0.5$ and $0<D<0.5$ are the non-seasonal and seasonal differencing parameters respectively, with additional constraints $0<d+D<0.5$ to assure the stationary of the process, $\phi(B)$, $\theta(B)$ are the non-seasonal polynomial operators with orders p, q respectively, $\Phi(B)$, $\Theta(B)$ are the seasonal polynomial operators with orders P, Q respectively:

$$\Phi(B)=1-\Phi_1 B-\Phi_2 B^2 - \text{L} - \Phi_P B^P,\quad \Theta(B)=1-\Theta_1 B-\Theta_2 B^2 - \text{L} - \Theta_Q B^Q.$$

For convenience, this paper is restricted to the SARFIMA model with $P=Q=0$, namely:

$$\phi(B)(1-B)^d\left(1-B^s\right)^D(X_t-\mu)=\theta(B)\varepsilon_t$$

and the ARFIMA model with $p=q=1$.

For the ARFIMA model, there are several methods chosen for its parameter estimation, including the Exact Maximum Likelihood method [8], WHI method [9] and Non-Linear Least Squares estimator [10]. For the SARFIMA model, Reisen et al. (2006) suggest a maximum likelihood method for its estimation [11]. However, this method is time-consuming when calculating the covariance matrix, especially for the high-frequency series with a relatively large sample size and when the AR coefficients, MA coefficients, seasonal and non-seasonal fractionally differencing parameters are all nonzero. Moreover, no further improvement for simplifying the procedure of this method, such as what Sowell (1991) does to improve the maximum likelihood estimation for the ARFIMA model, has yet been proposed. Bisognin & Lopes (2007) use the WHI method for the SARFIMA model's estimation in [12]. This method is simpler and faster in application. Because this paper discusses the forecasting of large sample high-frequency data using the SARFIMA model with nonzero AR and MA coefficients and non-seasonal fractionally differencing parameters, we use the WHI method for estimating the SARFIMA model.

For consistency, this paper also uses the WHI method to estimate the parameters of the ARFIMA. WHI is an approximated likelihood method. The discrete form of its likelihood function is given by:

$$L(\varsigma)=(2n)^{-1}\sum_{j=1}^{n-1}\left\{\ln f_X(\lambda_j,d)+\frac{I(\lambda_j)}{f_X(\lambda_j,\varsigma)}\right\}$$

where ς is the vector of unknown parameters $(d,D,\mu,\sigma_\varepsilon^2)'$, λ_j is the frequency, n is the sample size,

$f(\lambda_j,\theta)$ is the spectral density function of X_t and

$$I(\lambda_j)=(2\pi n)^{-1}\left|\sum_{t=1}^{n}X_t e^{it\lambda_j}\right|^2.$$

2.2. Short Memory Models

The two short memory models used in this paper are the ARMA model and the PAR model.

The ARMA model is formulated as:

$$\phi(B)(X_t - \mu) = \theta(B)\varepsilon_t.$$

By incorporating seasonal polynomial operators to the AR model, the PAR(p) model is defined as:

$$\phi(B)\Phi^s(B)(X_t - \mu) = \varepsilon_t$$

where s is the given period and $\Phi^s(B)$ is the seasonal polynomial operators with orders P:

$$\Phi^s(B) = 1 - \Phi_1 B^s - \Phi_2 B^{2s} - \text{L} - \Phi_P B^{Ps}.$$

This paper is restricted with $p = q = P = 1$, namely an ARMA (1,1) and a PAR(1). The parameters of the AR and PAR models are estimated by non-linear least squares method.

3. SIMULATION STUDY

To test the performance of different models for forecasting a periodic long memory series, we generate the SARFIMA $(p=1,d,q=1)\times(P=0,D,Q=0)_s$ artificial series X_t with zero mean and unit variance:

$$\phi(B)(1-B)^d(1-B^s)^D X_t = \theta(B)\varepsilon_t$$

$n \in \{300,1100\}$ when $s = 48$, $n \in \{500,2000\}$ when $s = 78$, autoregressive and moving average parameters $(\phi_1,\theta_1) = (0.3,0.4)$ and fractional differencing parameters $(d,D) = (0.2,0.2)$. Periods $s \in \{48,78\}$ are used to correspond to the periods of the intraday volume series examined in the next section. Consequently, there are four types of this artificial series. The last two periods of each series are left for forecasting; the former data are used for estimation. Forecasts are undertaken one-step in advance. For example, for {s = 48, n = 300} series, the 1st to 204th real data are used for estimation so that we can obtain the 205th predictive data. Then, the 2nd to 205 real data are used for estimation, and we forecast the 206th predictive data. Under each sample size, 100 duplicated series are generated to investigate the overall forecasting performance of different models. Figure 1 plots the examples of the last two periods of the four types of the artificial series. Figure 2 shows the ACF of these four series.

The periodicity does not seem to be apparent for all of these series, but the ACF shows significant autocorrelations both at the seasonal and non-seasonal lags for the two series. The Augmented Dickey-Fuller unit root test and two semiparametric tests, GPH test and the Gaussian Semi-parametric (GSP) test [13] [14] are also undertaken to examine their stationary status and long memory. Table 1 lists the ADF unit root test and the long memory test results. The ADF and long memory tests show that the two series are both stationary and with significant long memory.

For other duplicated series, their plots, ACF, stationarity and long memory properties are similar with the two series examined above. Due to space constraints, we provide only two examples here and do not elaborate on them.

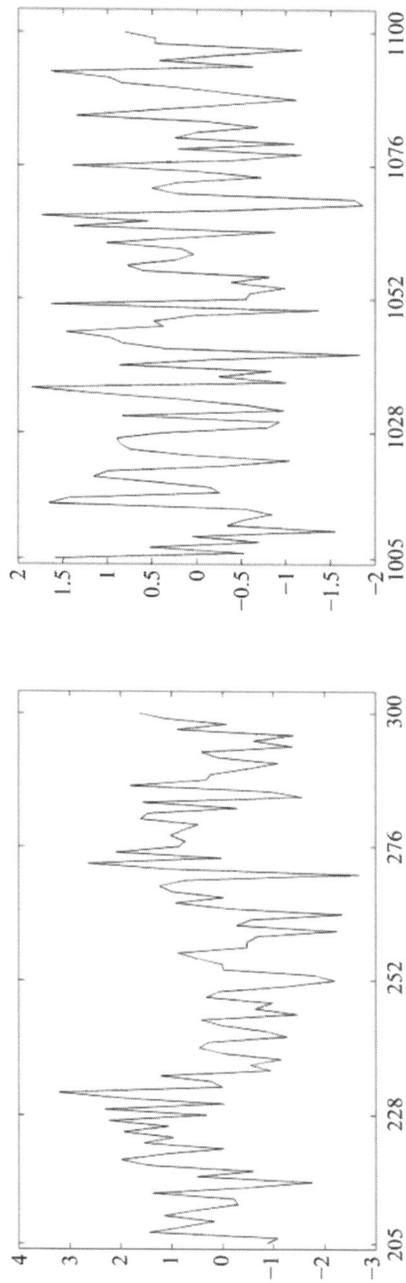

Figure 1. Data plots for the artificial series, the 205th to 300th data for n = 300 series (left), the 1005th to 1100th data for n = 1100 series (right).

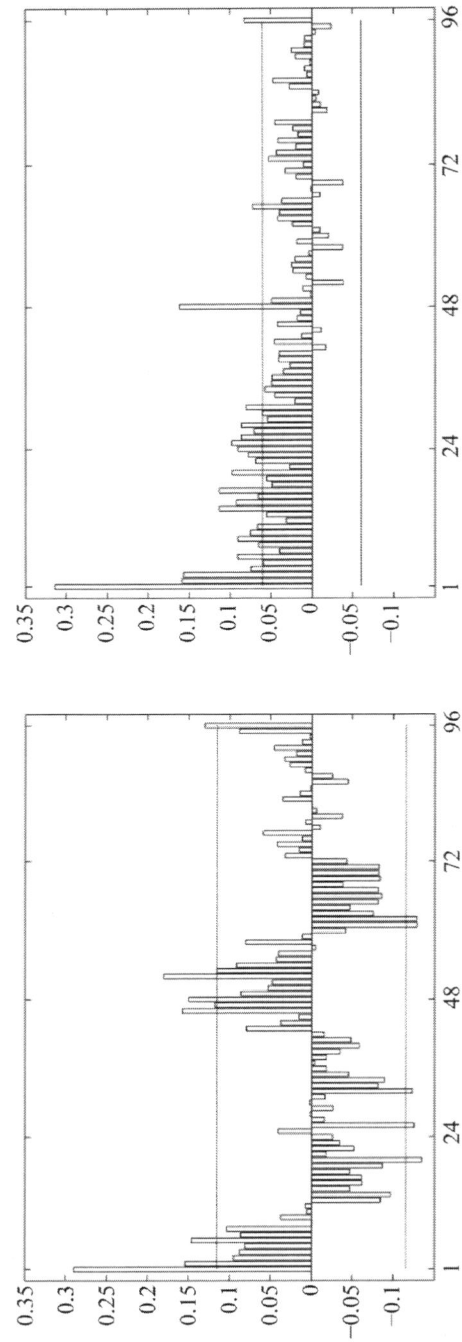

Figure 2. ACF for the artificial series, n = 300 series (left), n = 1100 series (right).

Table 1. ADF and long memory test results.

Test	$s = 48, n = 300$	$s = 48, n = 1100$	$s = 78, n = 500$	$s = 78, n = 2000$
ADF (t-statistics)	−12.7408*	−14.3546*	−15.3429*	−19.6710*
GPH (d value)	0.2264	0.2126	0.2435	0.2577
GSP (d value)	0.1798	0.2191	0.2657	0.2471

*Significant at 5% level.

The four models are then used to forecast the artificial series for testing their forecasting performances. Two sets of parameters are used for forecasting, especially for the SARFIMA model. One parameter is obtained by the WHI estimation. The other, as we already know the true parameters of the artificial SARFIMA series, is the set of parameters of the artificial series. Accordingly, we can take the performance of these two different parameter settings for the WHI estimation together to determine whether estimation bias would cause any negative effect. The statistical measure used in this paper for measuring forecasting accuracy is the mean squared error of the estimators (MSE), given by:

$$\text{MSE} = \sum_{t=1}^{k} \left(\hat{X}_t - X_t\right)^2$$

where k is the number of predicted data and X_t and $\hat{X}_t$ are the real and predictive value of the series respectively. For each type within the artificial series, we calculate the average MSE of the 100 duplicated series by:

$$\overline{\text{MSE}} = \sum_{i=1}^{n=100} \text{MSE} \Big/ 100.$$

Table 2 lists the average MSE of the four models for forecasting the four types of the artificial series. The SARFIMA model with known parameters is denoted as SARFIMA-known. The SARFIMA model with parameters estimated by WHI is denoted as SARFIMA-WHI. The averages of the MSE of these models for the 100 duplicated series are listed in the last row.

First, we can see from Table 2 that the PAR model performs best at forecasting all types of the artificial series. Its MSE are the smallest for most duplicated series. This finding indicates that a periodic short memory model is competent at predicting an SARFIMA series. The SARFIMA model using known parameters also performs well, with its average MSE ranked second. The non-periodic models, namely the ARFIMA and ARMA models, perform slightly worse than the periodic models. This finding indicates that considering periodicity is beneficial for accurately forecasting the artificial SARFIMA series, but the differences between the performances of the PAR, ARMA, SARFIMA-known and ARFIMA models are not very large. The differences of their average MSE are within 0.06.

However, the performance of SARFIMA-WHI falls significantly behind that of other models. Most of its MSE are much larger than that of others. Table 3 lists the average of the estimated parameters for the SAR- FIMA-WHI model.

We can see that the WHI method tends to underestimate both the seasonal and non-seasonal fractional differencing parameters d and D. This phenomenon is true especially for the n = 300 artificial series, for which the WHI method underestimate d and D by nearly 0.1. This finding indicates that the estimation bias is responsible for the loss of forecasting accuracy of the SARFIMA model using the WHI estimation.

4. EMPIRICAL STUDY

The intraday volume series is a very useful series for VWAP trading strategy, which splits and executes orders according to the predicted intraday volume distribution. Intraday volume series is a typical series that presents both periodicity and long memory. Here, we choose data gathered from the NASDAQ composite index and the

Table 2. The average MSEs for forecasting four types of the artificial series.

Type of the artificial series	ARFIMA	SARFIMA-known	SARFIMA-WHI	ARMA	PAR
$s = 48, n = 300$	1.0171	1.0123	1.2862	1.0404	0.9906
$s = 48, n = 1100$	1.0472	1.0223	1.3629	1.0630	1.0105
$s = 78, n = 500$	1.0370	1.0219	1.3460	1.0635	1.0117
$S = 78, n = 2000$	1.0585	1.0275	1.4514	1.0725	1.0144

Table 3. The average of the estimated parameters for the SARFIMA-WHI model.

Type of the artificial series	d	D	AR	MA
$s = 48, n = 300$	0.1354	0.0959	0.1078	0.1061
$s = 48, n = 1100$	0.1564	0.1626	−0.0304	0.1797
$s = 78, n = 500$	0.1822	0.1380	0.0292	0.1520
$s = 78, n = 2000$	0.1866	0.1823	0.1652	0.2838

Shanghai Stock Exchange 50 Index (SSE 50 Index)[1] to populate the sample for this description. We use two one-month samples. The SSE 50 Index ranges from January 4th to 31st 2011 in 5-minute intervals. The NASDAQ sample ranges from January 3rd to 31st 2011 in 5-minute intervals. Because the trading time for the Chinese stock market is 4 hours per day and for the American stock market is 6.5 hours per day, the total time of every trading day can be divided into 48 parts and 78 parts, respectively. Therefore, for 20 trading days, we obtained 960 and 1560 observed values from the Chinese market and American market, respectively. For each 5-minute interval, volume means the sum of all volumes traded within 5 minutes. Figure 3 shows the plots and the ACF of the sample series.

The periodicity and slow decay of the intraday volume series seem to be much more apparent than the artificial SARFIMA series. The plots show that the sample of the intraday volume series of the SSE 50 Index and NASDAQ composite index fluctuate in a U-shape and presents an apparent 48 and 78 periods, respectively. The ACF of the series show a very slow decay in the autocorrelation function both at the seasonal and non-seasonal lags for the series.

Next, we apply the four models to a one-year sample of the SSE 50 Index and NASDAQ composite index intraday volume to investigate their forecasting performance. This is an in-sample forecast comparison. The parameters of the models are estimated every month. The forecast is undertaken one-step ahead, using the monthly fixed parameters and historical rolling five-day data to forecast the next data. Table 4 and Table 5 list the statistics of the mean, maximum value, minimum value, ADF t-statistics and fractional differencing parameters d estimated by GPH and GSP of the SSE 50 Index and NASDAQ composite index intraday volume for each month in 2011, respectively.

On average, the maximum value of each month is more than 4 times the mean value and more than 28 times the minimum value.

This finding indicates a large deviation, partly due to the seasonal pattern. Although the ADF tests prove these series all to be stationary, most of the two semi-parametric estimators of the fractional differencing parameter, GPH and GSP, are near or above 0.5. These rates indicate that these stationary series have very strong long memories.

Applying the four models to the sample intraday volume series, we obtain the statistics of MSE of their forecasting, as listed in Table 6 and Table 7.

The results of the two tables are fairly similar. For most monthly samples, the ARFIMA model performs best, indicating that a fractional differencing is beneficial for forecasting intraday volume series. Meanwhile, the non- periodic models, the ARFIMA and the ARMA, seem to be superior to the periodic models. This finding indicates that, for forecasting intraday volume, adding periodicity may be unnecessary or redundant in terms of forecasting accuracy. The worst performance belongs to the SARFIMA model, of which the MSE are highest for forecasting all monthly samples. We can conclude that although this model is considered to be theoretically suitable for modeling periodic long memory series, it does not actually work very well on our intraday volume samples. Additionally, the two short memory models, the ARMA and PAR models, perform slightly worse than the ARFIMA model, but much better than the SARFIMA model does. This finding indicates that the short memory models are also competent in forecasting intraday volume.

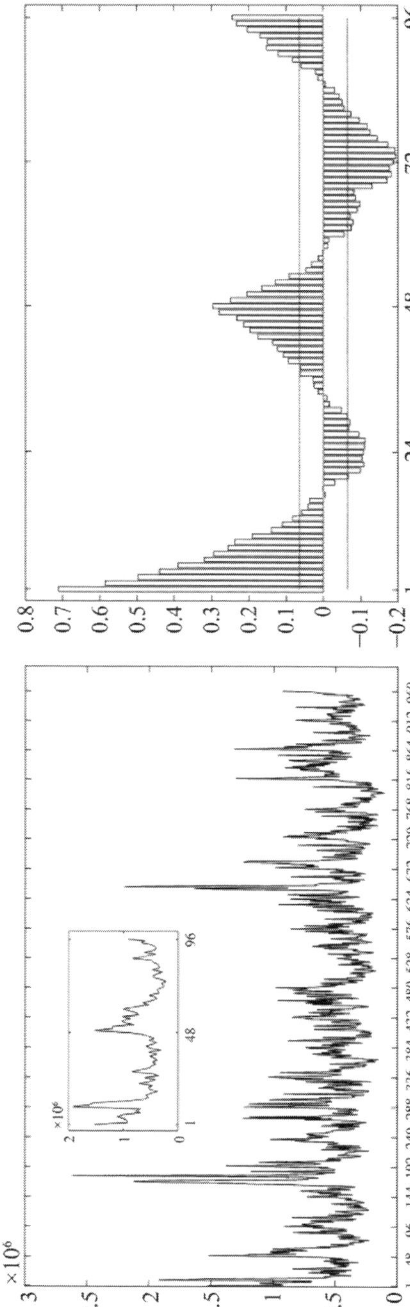

Figure 3. The plots and the ACF of the SSE 50 Index, January 4th to 31st 2011, 5-minute intervals.

Table 4. The statistics of the SSE 50 Index intraday volume for each month in 2011.

Month	Number of observations	Mean	Max	Min	ADF (*t*-statistics)	GPH (*d* value)	GSP (*d* value)
Jan.	960	517213	2617444	109180	−10.4575*	0.5885	0.5332
Feb.	720	578184	2032704	132651	−4.9658*	0.5653	0.5490
Mar.	1104	597276	2921257	134438	−8.7126*	0.5403	0.5735
Apr.	912	604845	3010937	175328	−8.6219*	0.5591	0.5880
May.	1008	315125	1052178	98968	−7.7034*	0.4277	0.4303
Jun.	1008	349801	1515968	85317	−6.6711*	0.4829	0.4470
Jul.	1008	412374	2153114	96716	−10.6532*	0.4949	0.5078
Aug.	1104	361702	2835370	83057	−9.1132*	0.5173	0.5241
Sep.	1008	269615	1563451	63966	−10.6051*	0.4033	0.4006
Oct.	768	386268	2983349	70697	−9.0214*	0.5066	0.4673
Nov.	1056	297883	1356324	66211	−6.5933*	0.5083	0.5431
Dec.	1056	247512	4117696	60367	−10.3629*	0.4056	0.4242
Average	-	-	-	-	-	0.5000	0.4990

*Significant at 5% level.

Table 5. The statistics of the NASDAQ intraday volume for each month in 2011.

Month	Number of observations	Mean	Max	Min	ADF (*t*-statistics)	GPH (*d* value)	GSP (*d* value)
Jan.	1560	55	195	0	−10.4543*	0.6201	0.6568
Feb.	1482	57	160	1	−8.3468*	0.5625	0.6194
Mar.	1794	67	206	3	−8.3348*	0.5373	0.6121
Apr.	1560	57	186	13	−8.9120*	0.5541	0.6630
May.	1638	62	179	5	−8.9724*	0.5117	0.6138
Jun.	1716	65	190	4	−8.5986*	0.4677	0.5749
Jul.	1560	66	183	1	−8.8776*	0.5265	0.5968
Aug.	1794	105	257	9	−6.2128*	0.6028	0.5980
Sep.	1638	97	225	29	−6.2058*	0.6183	0.6754
Oct.	1638	94	232	2	−5.9833*	0.6016	0.6858
Nov.	1638	78	186	1	−7.5528*	0.6128	0.6561
Dec.	1638	59	172	7	−7.2939*	0.5171	0.6365
Average	-	-	-	-	-	0.5610	0.6324

*Significant at 5% level.

Table 6. The MSE ($\times 10^{10}$) of the four models' forecasting results, SSE 50 Index.

Month	SARFIMA	ARFIMA	PAR	ARMA	Performance
Jan.	4.4344	2.9568	3.4543	3.1701	ARFIMA > ARMA > PAR > SARFIMA
Feb.	5.1260	3.5417	4.1332	3.8100	ARFIMA > ARMA > PAR > SARFIMA
Mar.	4.9637	3.6353	4.1220	3.8267	ARFIMA > ARMA > PAR > SARFIMA
Apr.	4.9465	3.6692	4.2262	3.8605	ARFIMA > ARMA > PAR > SARFIMA
May.	2.2025	1.0294	1.3206	1.0867	ARFIMA > ARMA > PAR > SARFIMA
Jun.	3.2072	1.5498	1.9334	1.6016	ARFIMA > ARMA > PAR > SARFIMA
Jul.	2.4091	1.6758	1.9425	1.8102	ARFIMA > ARMA > PAR > SARFIMA
Aug.	3.5710	1.7880	3.3168	3.6005	ARFIMA > PAR > ARMA > SARFIMA
Sep.	2.9869	1.7880	2.1075	1.9091	ARFIMA > ARMA > PAR > SARFIMA
Oct.	6.3777	3.4519	4.1498	3.6213	ARFIMA > ARMA > PAR > SARFIMA
Nov.	2.5193	1.1536	1.5092	1.1923	ARFIMA > ARMA > PAR > SARFIMA
Dec.	1.7676	0.8571	1.0541	1.0622	ARFIMA > PAR > ARMA > SARFIMA
Average	3.7093	2.2581	2.7725	2.5459	ARFIMA > ARMA > PAR > SARFIMA

In the column "Performance", ">" means to be superior to, e.g. ARFIMA > ARMA means the ARFIMA model is superior to the ARMA model for forecasting the corresponding months' intraday volume.

Table 7. The MSE ($\times 10^2$) of forecasting results, NASDAQ composite index.

Month	SARFIMA	ARFIMA	PAR	ARMA	Performance
Jan.	2.9006	2.2891	2.5360	2.3673	ARFIMA > ARMA > PAR > SARFIMA
Feb.	2.8684	2.0828	2.3794	2.1539	ARFIMA > ARMA > PAR > SARFIMA
Mar.	3.4062	2.6044	2.8700	2.6131	ARFIMA > ARMA > PAR > SARFIMA
Apr.	2.3652	2.1415	2.2669	2.1756	ARFIMA > ARMA > PAR > SARFIMA
May.	3.4284	2.4680	2.8457	2.4884	ARFIMA > ARMA > PAR > SARFIMA
Jun.	4.0275	2.6247	3.0821	2.5896	ARMA > ARFIMA > PAR > SARFIMA
Jul.	4.5266	3.4209	3.9391	3.4774	ARFIMA > ARMA > PAR > SARFIMA
Aug.	6.7548	4.5915	5.2309	4.5433	ARMA > ARFIMA > PAR > SARFIMA
Sep.	4.7220	3.5422	3.9096	3.4678	ARMA > ARFIMA > PAR > SARFIMA
Oct.	3.4851	2.6291	2.8474	2.5316	ARMA > ARFIMA > PAR > SARFIMA
Nov.	4.1874	3.0506	3.4239	3.0508	ARFIMA > ARMA > PAR > SARFIMA
Dec.	2.7012	2.0851	2.2770	2.0745	ARMA > ARFIMA > PAR > SARFIMA
Average	3.7811	2.7942	3.1340	2.7944	ARFIMA > ARMA > PAR > SARFIMA

In the column “Performance”, “>” means to be superior to, e.g. ARFIMA > ARMA means the ARFIMA model is superior to the ARMA model for forecasting the corresponding months’ intraday volume.

5. CONCLUSIONS

This paper evaluates the performance of the SARFIMA model at forecasting periodic long memory series, including the artificial SARFIMA series, the SSE 50 Index intraday volume series and NASDAQ composite index volume series. Three other models are also included in our study to compare their forecasting performances with that of the SARFIMA model.

For the artificial SARFIMA series, if we use the correct parameters for estimating and forecasting, it performs well relative to the other three models. However, if we use the parameters obtained by the WHI estimation, the forecasting performance of this model falls considerably behind other models. This phenomenon may be partly due to the estimation bias of the WHI estimation, which tends to underestimate both the seasonal and non-sea- sonal fractional differencing parameters. The PAR model performs best at forecasting all four artificial series. Meanwhile, the non-periodic models, namely the ARFIMA and ARMA models, do not perform as well as the periodic models. This outcome indicates that considering periodicity is beneficial for forecasting the artificial SARFIMA series.

For the intraday volume series, the ARFIMA model performs the best among all the models, indicating that fractional differencing is beneficial for forecasting the intraday volume series. For most monthly samples, the non-periodic models, the ARFIMA model and the ARMA model, seem to be superior to the periodic models. This outcome indicates that, for forecasting intraday volume, adding periodicity may be unnecessary or redundant in terms of forecasting accuracy. The SARFIMA model does not work very well on our intraday volume samples, exhibiting the worst performance among all the models used. In addition, the two short memory models, the ARMA and PAR models, also performed well compared to the SARFIMA model.

Handong Li, Xunyu Ye In summary, the SARFIMA was outperformed by other rival models in our study. Combining the results of the simulation and empirical study together, we conclude that the poor performance of the SARFIMA model may be caused by the inaccurate estimation obtained by the WHI method. The estimation method for this model still needs further improvement. Before more effective and more accurate estimation methods are proposed, we suggest that the SARFIMA model should be carefully applied when forecasting a high-frequency long memory time series with long periods.

REFERENCES

1. Engle, R. (2005) Handbook of Financial Econometrics. North Holland, Amsterdam.

2. Granger, C.W.J. and Joyeux, R. (1980) An Introduction to Long-Memory Time Series Models and Fractional Differencing. Journal of Time Series Analysis, 1, 15-29.

3. Hosking, J.R.M. (1981) Fractional Differencing. Biometrika, 68, 165-176.

4. Franses, P.H. and Ooms, M. (1997) A Periodic Long-Memory Model for Quarterly UK Inflation. International Journal of Forecasting, 13, 117-126.

5. Porter-Hudak, S. (1990) An Application of the Seasonal Fractionally Differenced Model to the Monetary Aggregates. American Statistical Association, 85, 338-344.

6. Ooms, M. and Franses, P.H. (2001) A Seasonal Periodic Long Memory Model for Monthly River Flows. Environmental Modeling & Software, 16, 559-569.

7. Nasr, A.B. and Trabelsi, A. (2005) Seasonal and Periodic Long Memory Models in the Inflation Rates. European Financial Management Association 2005 Annual Meetings, 31.

8. Sowell, F. (1992) Maximum Likelihood Estimation of Stationary Univariate Fractionally Integrated Time Series Models. Journal of Econometrics, 53, 165-188.

9. Whittle, P. (1953) Estimation and Information in Stationary Time Series. Arkiv för Matematik, 2, 423-434.

10. Doornik, J.A. and Ooms, M. (2004) Inference and Forecasting for ARFIMA Models, with an Application to US and UK Inflation. Studies in Nonlinear Dynamics and Econometrics, 8, 1218.

11. Reisen, V.A., Rodrigues, A.L. and Palma, W. (2006) Estimation of Seasonal Fractionally Integrated Processes. Computational Statistics & Data Analysis, 50, 568-582.

12. Bisognin, C. and Lopes, S.R.C. (2007) Estimating and Forecasting the Long-Memory Parameter in the Presence of Periodicity. Journal of Forecasting, 26, 405-427.

13. Geweke, J. and Porter-Hudak, S. (1983) The Estimation and Application of Long Memory Time Series Models. Journal of Time Series Analysis, 4, 221-238.

14. Robinson, P.M. (1995) Log Periodogram Regression of Time Series with Long Range Dependence. Annals of Statistics, 23, 1048-1072.

CHAPTER 5

Parallel and Hierarchical Mode Association Clustering with an R Package Modalclust

Yansong Cheng[1], Surajit Ray[2]

[1] Quantitative Science, Glaxo Smith Kline, King of Prussia, King of Prussia, Pennsylvania, USA

[2] School of Mathematics and Statistics, Glasgow University, Glasgow, UK

ABSTRACT

Modalclust is an R package which performs Hierarchical Mode Association Clustering (HMAC) along with its parallel implementation over several processors. Modal clustering techniques are especially designed to efficiently extract clusters in high dimensions with arbitrary density shapes. Further, clustering is performed over several resolutions and the results are summarized as a hierarchical tree, thus providing a model based multi resolution cluster analysis. Finally we implement a novel parallel implementation of HMAC which performs the clustering job over several processors thereby dramatically increasing the speed of clustering procedure especially for large data sets. This package also provides a number of functions for visualizing clusters in high dimensions, which can also be used with other clustering softwares.

KEYWORDS

Modality, Kernel Density Estimate, Mode, Clustering

1. INTRODUCTION

Cluster analysis is a ubiquitous technique in statistical analysis that has been widely used in multiple disciplines for many years. Historically cluster analysis techniques have been approached from either a fully parametric view, e.g. mixture model based clustering, or a distribution free approach, e.g. linkage based hierarchical clustering. While the parametric paradigm provides the inferential framework and accounts for the sampling variability, it often lacks the flexibility to accommodate complex clusters and are often not scalable to high dimensional data. On the other hand, the distribution free approaches are usually fast and capable of uncovering complex clusters by making use of different distance measures. However, the inferential framework is distinctly missing in the distribution free clustering techniques. Accordingly most clustering packages in R also fall under the two above mentioned groups of clustering techniques.

This paper describes a software program for cluster analysis that can knead the strengths of these two seemingly different approaches and develop a framework of parallel implementation for clustering techniques. For most model based approaches to clustering, the following limitations are well recognized in the literature: 1) the number of clusters has to be specified; 2) the mixing densities have to be specified, and as estimating the parameters of the mixture models is often computationally very expensive, we are often forced to limit our choices to simple distributions such as Gaussian; 3) computational speed is inadequate especially in high dimensions and this coupled with the complexity of the proposed model often limits the use of model-based techniques either theoretically or computationally; 4) it is not straightforward to extend model-based clustering to uncover heterogeneity at multiple resolutions, similar to the one offered by to the model free linkage based hierarchical clustering.

Influential work towards resolving the first three issues has been carried out in [1] -[7]. Many previous approaches have focused on model selection of mixtures by choosing the number of components, merging existing components or by determining the covariance structure of the mixture density under consideration, see [8] - [10]. They work efficiently if the underlying distribution is chosen correctly, but none of these model based approaches is designed to handle a completely arbitrary underlying distribution (see Figure 5 for one such example). That is, we think that limitations due to issues (3) and (4), above often necessitate the use of model- free techniques.

This paper describes a software program for cluster analysis that can knead the strengths of these two seemingly different approaches and develop a framework of parallel implementation for clustering techniques. The hierarchical mode association clustering—HMAC [11], which is constructed by first determining modes of the high-dimensional density and then associating sample points to those modes, is the first multivariate model based clustering approach resolving many of the drawbacks of standard model-based clustering. Specifically, it can accommodate flexible subpopulation structures at multiple resolutions while retaining the desired natural inferential framework of parametric mixtures. [12]

developed the inference procedure to the number of clusters of this approach. Modalclust is the package implemented in R (R Development Core Team, 2010) for carrying out HMAC along with its parallel implementation (PHMAC) over several processors. Though mode-counting or mode hunting has been extensively used as a clustering technique, most implementation are limited to univariate data. Generalization to higher dimensions was limited both due to the computational complexity of finding modes in higher dimension and the lack of any natural framework to study the inferential properties of modes in higher dimensions. The HMAC provides a computationally fast iterative algorithm for calculating the modes and thereby providing a clustering approach which is scalable to high dimensions. This article provides the description of the R package that implements HMAC and additionally provides an wide array visualization tools for representing clusters in high dimensions. Further, we propose a novel parallel implementation of the approach which dramatically reduces the computational time especially for large data sets, both in data dimen- sions and the number of observations.

This paper is organized as follows: Section 2 briefly introduces the algorithm of Modal Expectation Maximi- zation (MEM) and builds the notion of mode association clustering technique. Section 3 describes a parallel computing framework of HMAC along with computing time comparisons. Section 4 illustrates the implementation of clustering functions in the R package Modalclust along with examples of the plotting functions especially designed for objects of class hmac. Section 5 provides the conclusion and discussion. Comparison of Modal clustering with other popular model based and model free techniques are provided in the supplementary document.

2. MODAL EM AND HMAC

The main challenge for using mode-based clustering in high dimensions is the cost of computing modes, which are mathematically evaluated as local maximas of the density function with support on $\mathbb{R}^D$, D being the data dimension. Traditional techniques of finding local maxima, such as "hill climbing" works well for univariate data. But multivariate hill climbing is computationally expensive thereby limiting its use in high dimensions. [9] proposed an algorithm that solves a local maximum of a kernel density by ascending iterations starting from the data points. Since the algorithm is very similar to Expectation Maximization (EM) algorithm, it is named as

Modal Expectation Maximization (MEM). Define the mixture density as $f(x)=\sum_{i=1}^{K}\pi_i f_i(x)$. Now, given any initial value $x^{(0)}$, the MEM solves a local maximum of the mixture density by alternating the following two steps until it meets some user defined stopping criterion.

1. Let $p_i = \dfrac{\pi_i f_i\left(x^{(r)}\right)}{f\left(x^{(r)}\right)}, \quad i=1,\cdots,n$

2. Update $x^{(r+1)} = \text{argmax}_x \sum_{i=1}^{n} p_i \log f_i(x)$

Details of convergence of the MEM approach can be found in [11]. The above iterative steps provide a computationally simpler approach than grid search method for "hillclimbing" from any starting point $x \in \mathbb{R}^D$ by exploiting the properties of density functions.Given a multivariate kernel K, let the density of the data be given by

$$f(x|\Sigma)=\sum_{i=1}^{n}\frac{1}{n}K(x-x_i|\Sigma)$$

where Σ is the matrix of smoothing parameters. Further, in the special case of Gaussian kernels, i.e.,

$$K\left(x-x_i\middle|\Sigma\right)=\phi\left(x\middle|x_i,\Sigma\right)$$

where $\phi(\cdot)$ is the pdf of a Gaussian distribution, the update of $x^{(r+1)}$ is simply

$$x^{(r+1)}=\sum_{i=1}^{n}p_i x_i$$

allowing us to avoid the numerical optimization of Step 2.

Now we present the HMAC algorithm. First we scale the data and use a kernel density estimator, with a normal kernel to estimate the density of the data. The variance of the kernel, Σ is a diagonal matrix with all entries σ^2 denoted by $D(\sigma^2)$, thus σ^2 is the single smoothing parameter for all the dimensions. The choice of the smoothing parameter is an area of research in itself. In the present version of the program we incorporate the strategy of using pseudo degrees of freedom, proposed in [13]. Their strategy provides us with a range of smoothing parameters and exploring them from finest to coarsest resolution provides the user with the desired hierarchical clustering. First we describe the steps of Mode Association Clustering (MAC) for a single band- width σ^2.

1) Given a set of data $S=\{x_1,x_2,\cdots,x_n\}$, $x_i\in\mathbb{R}^d$ form kernel density

$$f\left(x\middle|S,\sigma^2\right)=\sum_{i=1}^{n}\frac{1}{n}\phi\left(x\middle|x_i,D\left(\sigma^2\right)\right) \tag{1.1}$$

2) Use $f\left(x\middle|S,\sigma^2\right)$ as the density function. Use each x_i, $i=1,2,\cdots,n$, as the initial value in the MEM algorithm and find one mode of

$f(x|S,\sigma^2)$ for each x_i. Let the mode identified by starting from x_i be $\mathcal{M}_\sigma(x_i)$.

3) Extract distinctive values from the set $\{\mathcal{M}_\sigma(x_i), i=1,2,\cdots,n\}$ to form a set G. Label the elements in G from 1 to $|G|$. In practice, due to finite precision, two modes are regarded equal if their distance is below a threshold κ. In our package, we use $\kappa = 10^{-4}$.

4) If $\mathcal{M}_\sigma(x_i)$ equals the kth element in G, x_i is put in the kth cluster.

We note that when the bandwidth σ increases, the kernel density estimator $f(x|S,\sigma^2)$ in (1.1) becomes smoother, and thus more points tend to climb to the same mode. This suggests a natural approach for hierar- chical organization (or "nesting") of our MAC clusters. Thus, given a range of bandwidths $\sigma_1 < \sigma_2 < \cdots < \sigma_L$, The clustering can be performed in the following bottom-up manner. Define the G_l as the collection of all the distinct modes obtained by MAC using the σ_l. First we perform MAC at the smallest bandwidth σ_1. At any bandwidth σ_l, the elements in G_{l-1} obtained from the preceding bandwidth are fed into MAC using the density $f(x|S,\sigma_l^2)$. The modes identified at this level form a new set of cluster is G_l. This procedure is repeated across all σ_l's. This preserves the hierarchy of clusters and thus the name Hierarchical Mode Association Clustering (HMAC). To summarize we present the HMAC procedure in the following box.

1) Start with the data $G_0 = \{x_1, \cdots, x_n\}$ and set level $l=0$ and initialize the mode association of the ith data point as $\mathcal{P}_0(x_i) = i$.

2) $l \leftarrow l+1$.

3) Form kernel density as in (1.1) using σ_l^2.

4) Cluster the elements in G_{l-1} by using density $f(x|S,\sigma_l^2)$. Let the set of distinct modes obtained be G_l.

5) If $P_{l-1}(x_i)=k$ and the kth element in G_{l-1} is clustered to the k'^{th} mode in G_l, then $P_l(x_i)=k'$. In another word, the cluster of x_i at level l is determined by its cluster representative in G_{l-1}.

6) Stop if $l=L$, otherwise go to Step 2.

3. PARALLEL HMAC

In this section we develop the method of parallel computing of HMAC (PHMAC) and its application together with some comparisons of performance of the parallel and non-parallel approach. The MAC approach is computationally expensive when the number of objects n becomes large. It requires that we use the MEM for each data point to find its local mamximum of the density. Note that for the HMAC, the steps for the level $l=2$ onwards only need to start the MEM from the modes of the previous level G_{l-1}, and hence the computational cost does not increase at the rate of n. Fortunately the MAC approach provides a natural framework for a “divide and conquer” clustering algorithm. One can simply divide the data into m partitions, perform modal clustering on each of those partitions, and pool the modes obtained from each of these partitions to form a collection G and apply the HMAC onward. If the user has access to several computing cores of the same machine or several processors of a shared memory computing cluster, the “divide and conquer” algorithm can be seamlessly parallelized. The PHMAC procedure is summarized as follows:

Step 1. Sphering transform the data $\mathbf{X}$ to form a new data set $\mathbf{Y}$.

Step 2. Let $G_0 = \{y_1, \cdots, y_n\}$. Divide the data ($n$ objects) into m partitions G_o^j randomly, $j = 1, 2, \cdots, m$.

Step 3. Perform HMAC on each of these subsets at the lowest resolution, i.e., using h_1 and get the modes G_1^j, $j = 1, 2, \cdots, m$.

Step 4. Pool the modes from each subset of data to form $G_1 = \bigcup_{j=1}^{m} G_1^j$.

Step 5. Perform HMAC starting from Step 2 and obtain the final hierarchical clustering.

Step 6. Transform $\mathbf{Y}$ back to $\mathbf{X}$.

Figure 1 shows one PHMAC example on the graph. In this figure, (a) shows the simulated data with four clusters along with the contour plot, where the color indicates the final clustering using PHMAC; (b) shows the four random partitions of the unlabeled data along with the modes (red asterisks) at each partition; (c) shows the mode obtained from the four partitions; (d) shows the final modes (green triangles) starting from the modes of the partitioned data. A demonstration of different steps of parallel clustering with four random partitions is given in Figure 1. The original data set is partitioned into 4 random subsets, and initial modal clustering is performed within the partitions. In the next step, the modes of each of these partitions are merged to form the overall modal clusters in Figure 1(c).

Modes have a natural hierarchy and it is computationally easy to merge modes from different partitions. In practice, we need to decide the best choice of the partition and how many partitions to use. In this section, we provide some guidelines regarding the choices, without exploring their quality in details. In the absence of any other knowledge, one should randomly partition the data. Other choices include partitioning data based on certain coordinates which form a natural clustering, and then taking products of a few of those coordinates to build the overall partition. This strategy might increase the computational speed by restricting the modes within a relatively homogeneous set of observations. Another choice might be to sample the data and build partitions based on the modes of the sampled data.

The PHMAC we proposed uses parallel computing at the first level of HMAC and then use non-parallel computing from the second level onwards. Therefore, the number of partitions to minimize the computational time is a complex function of the number of available processors, the number of observations and the bandwidth parameter of the KDE. If one uses too many partitions, one might speed up the first step, but would have the risk of ending up with too many modes for the next level, where the hill climbing is done from the collection of modes from each partition with respect to the overall density. In contrast, for a large n, if one chooses too few partitions or no partitions, this would lead to a huge computational cost at the first step. Moverover, the choice of the smoothing parameter will also determine how many modes one needs to start from at the merged level.

We compare the computing speed of parallel versus serial clustering using 1, 2, 4, 8 and 12 multi-core processors. Tests were performed on a 64 bits 4 Quad Core AMD 8384 (2.7 Ghz each core), with 16 GB RAM running Linux Centos 5 and R version 2.11.0 From Table 1, it is clear to observe that parallel computing significantly increases the computing speed. Because the KDE is a sum of kernels centered at every data point, the amount of computation needed to identify the mode associated with a single point grows linearly with n. The computational complexity of clustering all the data by MAC is thus quadratic in n. Suppose we have p processors, then the computing complexity for the MAC is n^2 and for parallel computing of MAC is thus $(n/p)^2$. However, as discussed before, we can see that the computational speed is not a monotone decreasing function of the number of processors. Theoretically, it is true that more processors can reduce the computing complexity at the initial step. However, in practice, if the data set is not sufficiently large, using more processors may not save time, as it may produce a large number of modes for the next level of HMAC. When the $n = 10{,}000$ or $n = 50{,}000$, including more processors provides a dramatic decrease in computing time, whereas for $n = 2{,}000$, there is no clear decrease in time elapsed when using 4 or 8 processors instead of the maximum 12 processors. For $n = 50{,}000$, the decrease in computing time from 1 processor to using 12 processors is more than 40 fold (see Figure 2), but even if the user is able to use just two processors, the computing time is reduced to 1/3 of how long a single processor would take. Even for $n = 20{,}000$, the advantage of using 12 processors is almost 30 fold, whereas for $n = 2{,}000$, the advantage is only 8 folds. In fact, the lowest time is actually clocked by 8 processors for $n = 20{,}000$, but using all 12 processors does not increase the time significantly. These comparisons show the potential for parallelizing the modal clustering algorithm and its inherent use for clustering high throughput data.

The R package Modalclust was created to implement the HMAC and PHMAC. There are also some plotting tools that give the user a comprehensive visual and understanding of the clustering result. Sources, binaries and documentation of Modalclust are available for download from the Comprehensive R Archive Network http://cran.r-project.org/ under the GNU Public License.

4. EXAMPLE OF USING R PACKAGE MODALCLUST

In this section, we demonstrate the usage of the functions and plotting tools that are available in the Modalclust package.

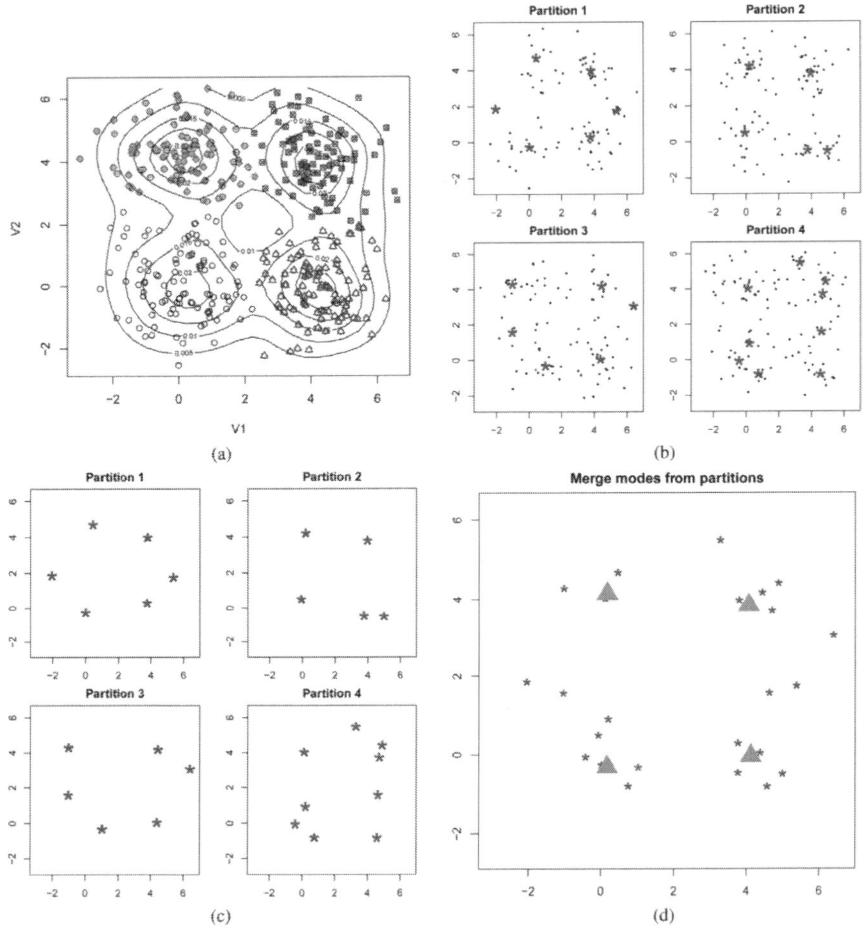

Figure 1. Steps in parallel HMAC procedure for a simulated data set.

4.1. Modal Clustering

First, we provide an example of performing modal clustering to extract the subpopulations in the logcta20 data. The description of the dataset is given in the package. The scatter plot, along with its smooth density, is provided in Figure 3. First, we use the following command to download and install the package:

Table 1. Comparison of computing time (elapsed time in seconds) using different number of processors.

Data dimensions		Number of processors				
		1	2	4	8	12
$n = 2000$	$d = 2$	56.58	17.01	7.84	6.91	8.02
$n = 2000$	$d = 20$	323.16	128.13	112.42	190.11	250.22
$n = 2000$	$d = 40$	730.18	560.16	687.79	764.29	753.36
$n = 10,000$	$d = 2$	3849.83	871.33	276.88	145.61	131.22
$n = 10,000$	$d = 20$	8410.96	1694.82	585.33	536.32	459.88
$n = 50,000$	$d = 2$	210295.29	71152.82	23383.61	11959.24	4875.64

R > install.packages ("Modalclust")

R > library ("Modalclust")

Using the following command, we can get the standard (serial) HMAC and parallel HMAC using two pro- cessors for logctA20 data.

R > logcta20.hmac < −phmac(logcta20,npart=1,parallel=FALSE)

R > logcta20p2.hmac < −phmac(logcta20,npart=2,parallel=TRUE)

Both implementation results are given in Figure 4, which clearly identifies the three distinct subpopulations. Other model-based clustering methods, such as EM-clustering or K-means, could not capture the subpopulation structure, as the individual subpopulation is not a normal density. Distance based clustering method e.g., hierarchical clustering, with a range of linkage functions performed even worse.

By default, the function selects an interesting range of smoothing parameters with ten σ^2 values, and the final clustering only shows the results from the levels which produced merging from the previous level. For example, for the logcta20, the smoothing parameters chosen automatically are

R > logcta20.hmac$sigma

[1] 0.26 0.29 0.31 0.34 0.38 0.43 0.49 0.58 0.72 0.94,

which are chosen using the spectral degrees of freedom criterion introduced in [10]. Though we started with 10 different smoothing levels, the final clustering shows only 6 different levels along with a decreasing number of hierarchical cluster.

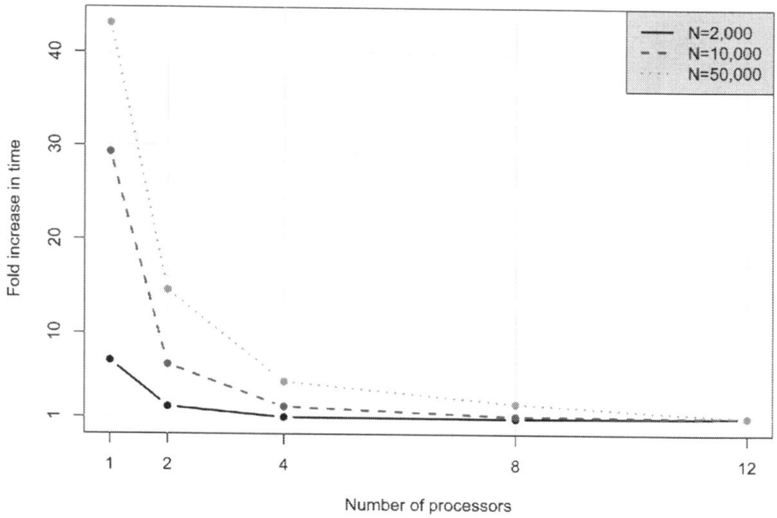

Figure 2. Comparison of fold increase in time for clustering two dimensional data of different sample sizes with respect to using 12 processors.

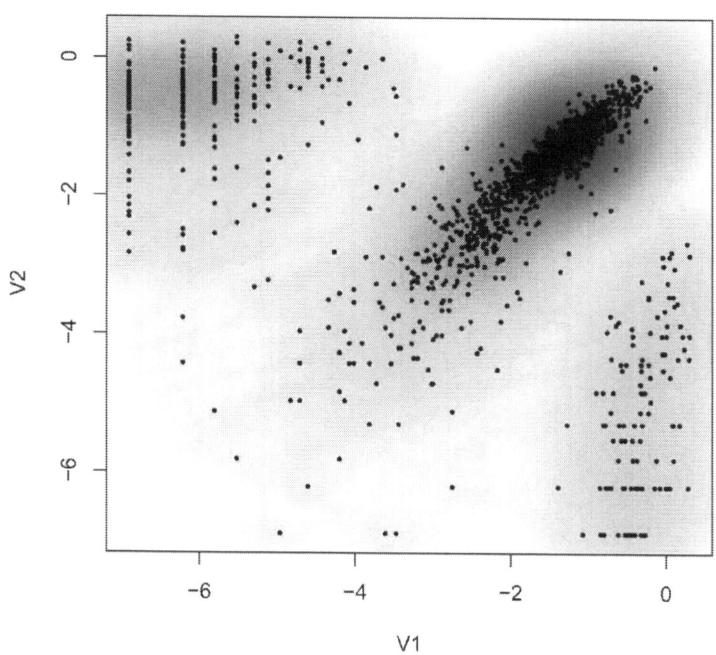

Figure 3. Smoothing scatter plot of logctA20 data.

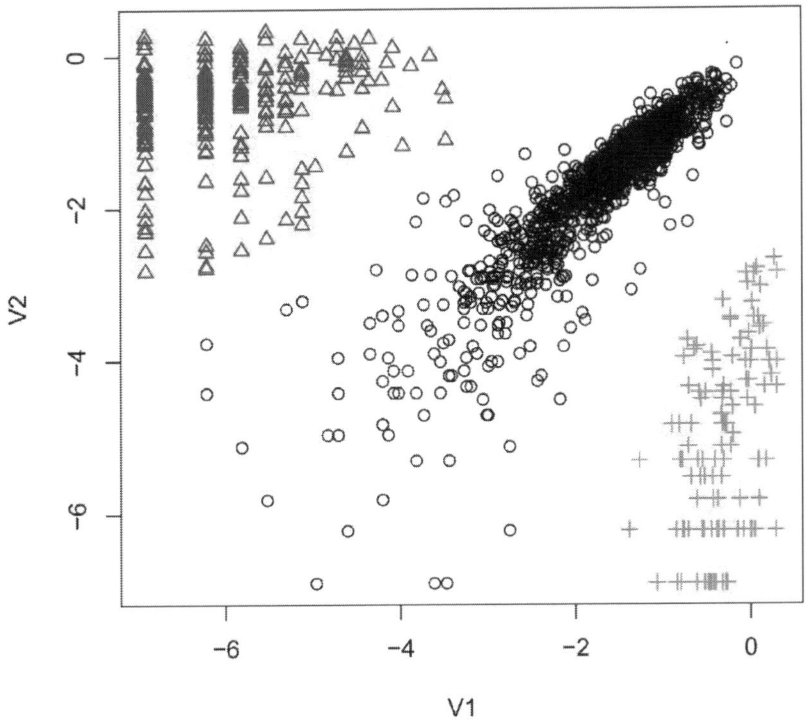

Figure 4. HMAC output of logctA20 data.

```
R > logcta20.hmac$level

[1] 1 2 3 3 3 4 4 4 5 6

R > logcta20.hmac$n.cluster

[1] 11 7 5 5 5 3 3 3 2 1
```

The user can also provide smoothing levels using the option sigmaselect in phmac. There is also the option of starting the algorithm from user defined modes instead of the original data points. This option becomes handy if the user wishes to merge clusters obtained from other clustering methods, e.g., EM-clustering or K-means.

4.2. Some Examples of Plotting

There are several plotting functions in Modalclust, which can be used to visualize the output from the function phmac. The plotting functions are defined on object class hmac, which is the default class of a phmac output. These plot functions will be illustrated through a data set named disc2d, which has 400 observations displaying the shape of two half discs. The scatter plot of disc2d along with its contour plot are given in Figure 5.

First, we introduce the standard plot function for an object of class "hmac". This unique and informative plot shows the hierarchical tree obtained from modal clustering. It can be obtained by

R > data ("disc2d.hmac")

R > plot (disc2d.hmac)

The dendrogram obtained from the disc2d data is given in Figure 6. The y-axis gives the different levels, and the tree displays the merging at different levels. There are several options available for drawing the tree, including starting the tree from a specific level, drawing the tree only up to a desired number of clusters, and comparing the clustering results with user defined clusters.

There are some other plotting functions that are designed mainly for visualizing clustering results for two dimensional data, although one can provide multivariate extensions of the functions by considering all possible pairwise dimensions. One can obtain the hard clustering of the data for each level using the command

R > hard.hmac(disc2d.hmac)

Alternatively, the user can specify the hierarchical level or the number of desired clusters, and obtain the corresponding cluster membership (hard clustering) of the data. For example, the plot inFigure 7 can be obtained by either of the following two commands:

R > hard.hmac (disc2d.hmac, n.cluster=2)

R > hard.hmac (disc2d.hmac, level=3)

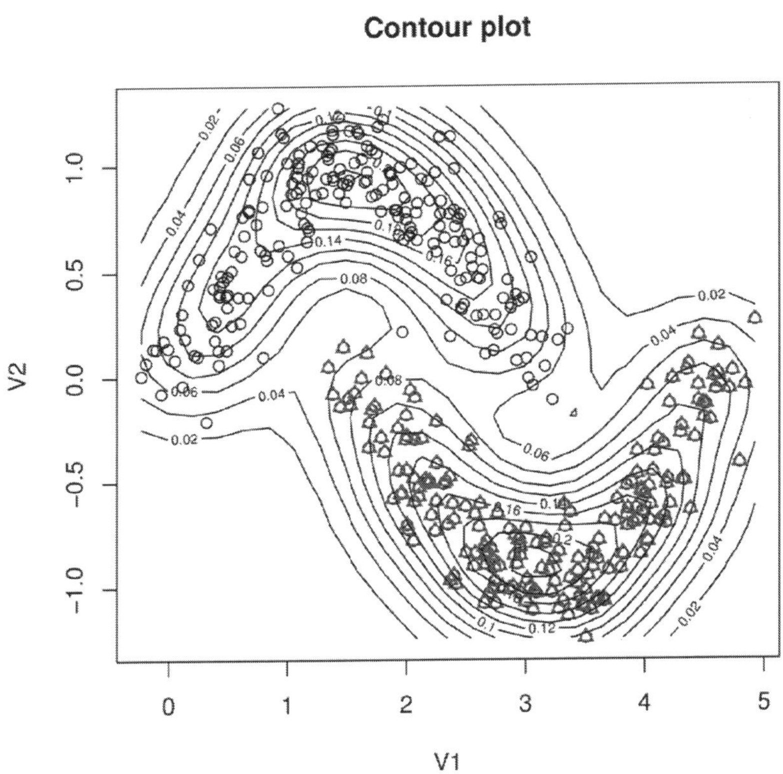

Figure 5. The scatter plot of disc2d data along with its probability contours.

Another function, which allows the user to visualize the soft clustering of the data, is based on the posterior probabilities of each observation belonging to the clusters at a specified level. For example, the plot in Figure 8 can be obtained using

R > soft.hmac (disc2d.hmac, n.cluster=3)

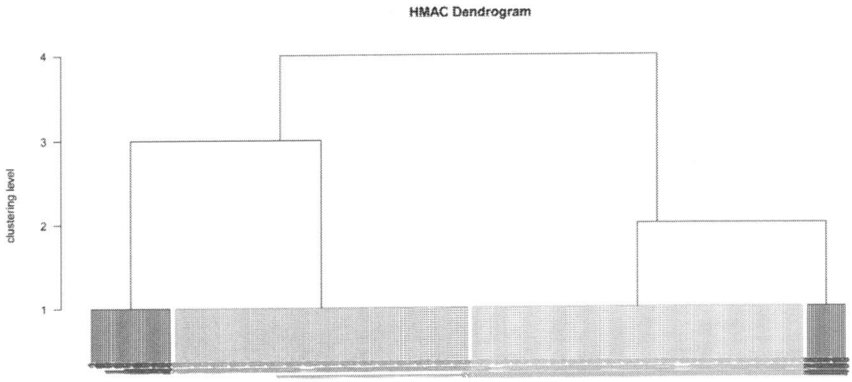

Figure 6. Hierarchical tree (Dendrogram) of disc2d data showing the clustering at four levels of smoothing.

The plot enables us to visualize the probabilistic clustering of the three cluster model. A user can specify a probability threshold for assigning observations which clearly belong to a cluster or lie in the "boundary" of more than one cluster. Points having posterior probability below the user specified boundlevel (default value 0.4) are assigned as boundary points and colored in gray. InFigure 8, we have five boundary points among the 400 original observations. Additionally, at any specified level or cluster size, the plot=FALSE option in hard.hmac returns the cluster membership. Similarly, plot=FALSE option in soft.hmac returns a list that contains the posterior probability of each observation and boundary points.

R > disc2d.2clust < −hard.hmac (disc2d.hmac,n.cluster=2, plot=FALSE)

R > disc2d.2clust.soft < −soft.hmac (disc2d.hmac,n.cluster=2, plot=FALSE)

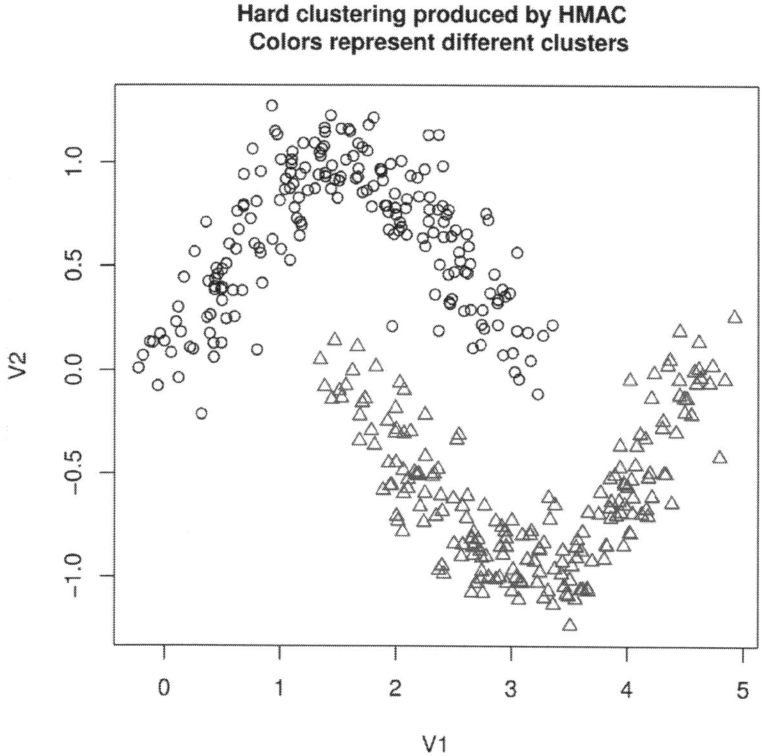

Figure 7. Hard clustering for disc2d data at level 3.

5. DISCUSSION

Modalclust performs a hierarchical model based clustering allowing for arbitrary density shapes. Parallel computing can dramatically increase the computing speed by splitting the data and running the HMAC simul- taneously on multi-core processors. Plotting functions give the user a comprehensive visualizing and under- standing of the clustering result. One future work from this stage would be to increase computing speed, especially for large data set. From the discussion in Section 3, it is clear to see, parallel computing increases the computing speed a lot. That relies on the computing equipment. If one user has no multicore or a few multicore processors available, it will take a lot of the computing resources when clustering large data sets. One potential way to solve the computing speed problem is using k-means or other faster clustering techniques initially, and using the HMAC from the centers of each cluster of initial clustering results. For example, if we have a data set with 20,000 observations, we can use k-means clustering and choose a certain number of centers, like 200 centers and run k-means clustering first. And then we start from the centers of 200 clusters and clustering by HMAC. Theoretically it is a sub-optimal way compared with running HMAC for all points. In practice, it is very useful to reduce the computing costs and still obtain the right clustering.

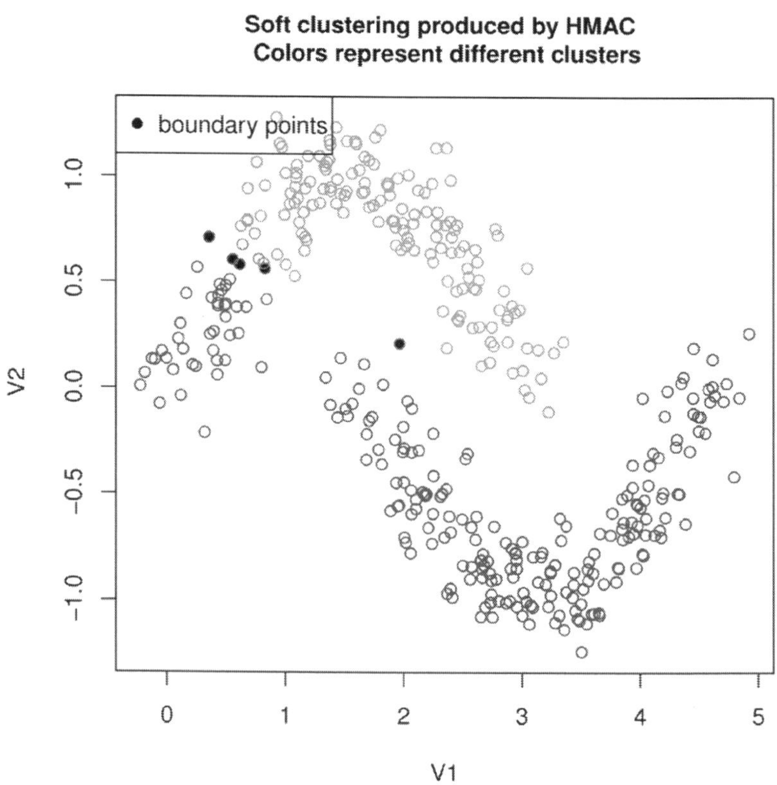

Figure 8. Soft clustering for disc2d data at level 2.

In addition, we are currently working on an implementation of modal clustering for online or streaming data, where the goal would be to update an existing cluster with the new data without storing all the original data points and allowing for creation of new clusters and merging of existing clusters.

Sources, binaries and documentation of Modalclust are available for download from the Comprehensive R Archive Network http://cran.r-project.org/ under the GNU Public License.

REFERENCES

1. Fraley, C. (1998) Algorithms for Model-Based Gaussian Hierarchical Clustering. SIAM Journal on Scientific Computing, 20, 270-281.

2. Fraley, C. and Raftery, A. (1998) How Many Clusters? Which clustering method? Answers via Model-Based Cluster Analysis. The Computer Journal, 41, 578-588.

3. Fraley, C. and Raftery, A. (1999) Mclust: Software for Model-Based Cluster Analysis. Journal of Classification, 16, 297-306.

4. Fraley, C. and Raftery, A. (2002) Model-Based Clustering, Discriminant Analysis, and Density Estimation. Journal of the American Statistical Association, 97, 611-631.

5. Fraley, C. and Raftery, A. (2002) Mclust: Software for Model-Based Clustering, Density Estimation and Discriminant Analysis. Tech. Rep., DTIC Document.

6. Ray, S. and Lindsay, B. (2005) The Topography of Multivariate Normal Mixtures. The Annals of Statistics, 33, 2042-2065.

7. Ray, S. and Lindsay, B. (2007) Model Selection in High Dimensions: A Quadratic-Risk-Based Approach. Journal of the Royal Statistical Society: Series B (Statistical Methodology), 70, 95-118.

8. Baudry, J., Raftery, A., Celeux, G., Lo, K. and Gottardo, R. (2010) Combining Mixture Components for Clustering. Journal of Computational and Graphical Statistics, 19, 332-353.

9. Hennig, C. (2010) Methods for Merging Gaussian Mixture Components. Advances in Data Analysis and Classification, 4, 3-34.

10. Tantrum, J., Murua, A. and Stuetzle, W. (2003) Assessment and Pruning of Hierarchical Model Based Clustering. Proceedings of the 9th ACM SIGKDD International Conference on Knowledge Discovery and Data Mining, 197-205.

11. Li, J., Ray, S. and Lindsay, B. (2007) A Nonparametric Statistical Approach to Clustering via Mode Identification. Journal of Machine Learning Research, 8, 1687-1723.

12. Cheng, Y. and Ray, S. (2014) Multivariate Modality Inference Using Gaussian Kernel. Open Journal of Statistics, 4, 419-434.

13. Lindsay, B., Markatou, M., Ray, S., Yang, K. and Chen, S. (2008) Quadratic Distances on Probabilities: A Unified Foundation. The Annals of Statistics, 36, 983-1006.

CHAPTER 6

On Detecting Sudden Changes in the Unconditional Volatility of a Time Series

Dilip Kumar

Indian Institute of Management, Kashipur, India

ABSTRACT

The present study highlights the drawback of using Sanso, Arago and Carrion's (2004) AIT-ICSS algorithm in detecting sudden changes in the unconditional volatility when long memory is present in volatility. Simulation experiments show that the AIT-ICSS test is severely oversized and exhibits low power when long memory is present in volatility.

KEYWORDS

AIT-ICSS Algorithm, Long Memory, Sudden Change, Volatility

1. INTRODUCTION

Volatility modeling and forecasting play a crucial role in financial market and have been well-researched in the area of economics and finance because of its importance in capital market theories. Volatility of asset returns highlights the risk or uncertainty associated with the asset and, hence, exploring the behavior of volatility of asset returns is relevant for the pricing of financial assets, risk management, portfolio selection, trading strategies and the pricing of derivative instruments [1]. Volatility of asset returns does not remain constant and vary over time; however, sometime we observe sudden changes in volatility of returns. Macroeconomic and political shocks may result in sudden changes in the volatility of asset returns. Such sudden changes in volatility in turn influence the intensity and the direction of information flow across markets, stocks or portfolios as shown by Ross [2]. Hence, detecting sudden changes in the unconditional variance is an important issue in finance, because it is well known that volatility does not stay constant over time. Various approaches are available to detect sudden changes in unconditional volatility. Among them, Inclan and Tiao's [3] Iterated Cumulative Sum of Squares (IT-ICSS) and Sanso, Arago and Carrion's [4] (AIT-ICSS) algorithms are more popular. The AIT-ICSS algorithm deals with the shortcomings of the IT-ICSS algorithm by taking into account conditional heteroskedasticity that is commonly observed in financial time series and also, the serial correlation in the conditional volatility series. It has met with a great deal of success empirically in being able to identify the breakpoints in volatility regimes [5] - [7].

However, in some instances, it has been applied even when long memory is present in the conditional volatility series, such as by Malik et al. [5], Hammoudeh and Li [6], Wang and Moore [7]. This strikes us as being possibly inappropriate because a careful look at the theory behind the AIT-ICSS algorithm reveals that this procedure is designed to deal only with short term dependence in the conditional volatility series. Hence, it becomes an interesting question to investigate how well the AIT-ICSS algorithm performs under long memory. Therefore, we undertake a simulation experiment to assess the size and power properties of the AIT-ICSS test when long memory is present in volatility using data generating processes from various specifications of the fractionally integrated generalized autoregressive conditional heteroskedasticity (FIGARCH) model. What our results show is that indeed the AIT-ICSS test performs badly under long memory in the sense of being badly oversized. On the application side, we select four major benchmark indices, which are S&P 500, FTSE 100, Nikkei 225 and CAC 40. We first estimate the fractional integration parameter of these indices using the FIGARCH model and find evidence of long memory in the conditional volatility of these series. We then detect sudden breaks in the unconditional volatility of these series using the AIT-ICSS algorithm and find that most of the detected breaks are spurious in nature, which can be explained based on the evidence from our simulation experiments.

The remainder of this paper is organized as follows: Section 2 introduces the AIT-ICSS algorithm and the associated problem when long memory is incorporated in the series. In Section 3, we undertake Monte Carlo simulation experiments to assess the AIT-ICSS algorithm. Section 4 describes the problem of AIT-ICSS when applied on data and Section 5 concludes with a summary of our main findings.

2. METHODOLOGY

In the AIT-ICSS test, suppose $\varepsilon_t \sim iid\left(0,\sigma^2\right)$. Suppose the variance within each interval is given by τ_j^2, where $j=0,1,\cdots,N_T$ and N_T is the total number of variance changes in T observations, and $1 < k_1 < k_2 < \cdots < k_{NT} < T$ are the change points.

$$\sigma_t^2 = \tau_0^2 \quad \text{for } 1 < t < \kappa_1 \tag{1.1}$$

$$\sigma_t^2 = \tau_1^2 \quad \text{for } \kappa_1 < t < \kappa_2 \tag{1.2}$$

$$\sigma_t^2 = \tau_{NT}^2 \quad \text{for } \kappa_{NT} < t < T \tag{1.3}$$

A cumulative sum of squares procedure is used to estimate the number of change points and is given as:

$$C_k = \sum_{t=1}^{k} \varepsilon_t^2 \tag{2}$$

where $k = 1, \cdots, T$. The AIT test statistic to detect sudden change is given as:

$$AIT = \max_k \sqrt{T}\,|G_k| \tag{3}$$

where

$$G_k = \frac{1}{\sqrt{\hat{\lambda}}}\left[C_k - \left(\frac{k}{T}\right)C_T\right],\ k = 1, \cdots, T \tag{4.1}$$

where C_T is the sum of squared residuals from the whole sample period.

$$\hat{\lambda} = \hat{\gamma}_0 + 2\sum_{i=1}^{m}\left[1 - \frac{i}{m+1}\right]\hat{\gamma}_i \tag{4.2}$$

$$\hat{\gamma}_i = \frac{1}{T}\sum_{t=i+1}^{T}\left(\varepsilon_t^2 - \hat{\sigma}^2\right)\left(\varepsilon_{t-i}^2 - \hat{\sigma}^2\right), \text{ and } \hat{\sigma}^2 = \frac{1}{T}C_T$$

It needs to be noted that when we make use of Equation (4.2) we are implicitly assuming that the spectral density at zero of the squared returns is well behaved and, in particular, that as m gets large, the right hand side of Equation (4.2) converges to the following:

$$S(0) = \gamma_0 + 2\sum_{j=1}^{\infty} \gamma_j \tag{5}$$

In practice, the lag truncation parameter m in Equation (4.2) is estimated using the procedure given in Newey and West [8]. The innovation in the AIT-ICSS algorithm with respect to the IT-ICSS algorithm is related to incorporating the correction for conditional heteroskedasticity and also for the possible serial correlation in conditional volatility. However, in Equation (4.2), if $m \to \infty$, $\hat{\lambda}$ will be well behaved only when there is short-term dependence in the conditional volatility series which rules out the presence of long memory in volatility. This is not an ancillary aspect of the AIT-ICSS algorithm, but the essential concept that distinguishes it from the original IT-ICSS algorithm.

3. MONTE CARLO SIMULATION EXPERIMENT

This section presents the performance of the AIT-ICSS algorithm in the presence of long memory in conditional volatility via simulation. The sample size, the number of Monte Carlo trials and the significance level are taken to be (T = 100, T = 200, T = 500 and T = 1000), 10,000 and 5%[1]respectively. We consider different specifications of FIGARCH (p, d, q) (i.e., FIGARCH (0, d, 0), FIGARCH (1, d, 0) and FIGARCH (1, d, 1)) models. The FIGARCH (p, d, q) model is given as:

$$\varepsilon_t = \sqrt{h_t}u_t;\ \phi(L)(1-L)^d\,\varepsilon_t^2 = \omega + [1-\beta(L)]v_t,\ u_t \sim iidN(0,1) \tag{6}$$

where $v_t = \varepsilon_t^2 - h_t$ and $\phi(L) = [1 - \alpha(L) - \beta(L)]$.

We consider the following cases under (6).

FIGARCH (1, d, 1):

$$\begin{cases} (\omega, \phi, d, (\beta)) = (0.1, 0.975, 0, (0.1, 0.3, 0.5, 0.7)) \\ (\omega, \phi, d, (\beta)) = (0.1, 0.25, 0.5, (0.1, 0.3, 0.5, 0.7)) \\ (\omega, \phi, d, (\beta)) = (0.1, 0.05, 0.75, (0.1, 0.3, 0.5, 0.7)) \end{cases}$$

FIGARCH (1, d, 0):

$$\begin{cases} (\omega, \phi, d, (\beta)) = (0.1, 0, 0.5, (0.1, 0.3, 0.45)) \\ (\omega, \phi, d, (\beta)) = (0.1, 0, 0.75, (0.1, 0.3, 0.5, 0.7)) \end{cases}$$

FIGARCH (0, d, 0)

$$(\omega, \phi, (d), \beta) = (0.1, 0, (0, 0.25, 0.50, 0.75), 0)$$

Under the null hypothesis, there is no sudden change in σ_t^2. These specifications can help us to examine the size of the AIT test. For the case of FIGARCH (1, d, 0) model, one specification contains β up until 0.45 because for the FIGARCH model, $(\phi + d - \beta)$ should be non-negative. Under the alternative hypothesis, there exists a sudden change in volatility; hence, we incorporate a break at the 50th percentile[2] of the series to compute the power of the test. To generate ε_t^2 for assessing the power of the test, we first generate e_t using the respective FIGARCH specification and keep first part of the series as it is and multiply the second half by (1 + l), where l indicates the percentage change in the unconditional volatility of the series. Here the results for FIGARCH

(1, d, 1) and FIGARCH (1, d, 0) specifications are very similar and due to lack of space we only report the results of all FIGARCH (1, d, 1) specifications. We also report the power of the test for FIGARCH (0, d, 0) for different values of d (0, 0.5 and 0.75).

Table 1 presents the results of size of the test for different specifications of the FIGARCH (p, d, q) model. The results indicate that the AIT-ICSS test is severely oversized for all the cases with the exception of FIGARCH (0, d, 0) model and FIGARCH (1, d, 1) model (for $\beta = 0.10$) when $d = 0$.

Table 2 reports the power of the test. The FIGARCH (0, d, 0) model with $d = 0$ exhibits good power for a sudden change in the unconditional volatility of 100% or more and the power of the test increases as sample size increases even for sudden change of 30% or more. On the other hand, for non-zero d, we observe good power for other specifications only when there is sudden change of 300% or more for sample size greater than 200. The FIGARCH (0, 0.75, 0) model exhibits the worst power of all the specifications under study.

This indicates that when long memory is present in the conditional volatility of a time series, the AIT-ICSS algorithm breaks down and is no longer suitable for detecting sudden changes in its unconditional volatility.

4. APPLICATION

We apply the AIT-ICSS algorithm on weekly data of four major benchmark series having long memory in the conditional volatility. The data period is from January 1996 to March 2015. Table 3reports the estimated values of long memory parameter d using the FIGARCH (1, d, 1) specification. The results clearly show that there exists long memory in the conditional volatility of all the series under study.

Table 1. Size of the test.

FIGARCH (1, d, 1) ($d=0$ & $\phi=0.975$)					FIGARCH (0, d, 0) for different values of d				
β	$T=100$	$T=200$	$T=500$	$T=1000$	d	$T=100$	$T=200$	$T=500$	$T=1000$
0.10	0.044	0.042	0.052	0.055	0.00	0.017	0.024	0.033	0.035
0.30	0.095	0.088	0.113	0.115	0.25	0.106	0.179	0.332	0.464
0.50	0.191	0.202	0.266	0.263	0.50	0.149	0.256	0.484	0.629
0.70	0.275	0.403	0.489	0.541	0.75	0.103	0.170	0.253	0.303
FIGARCH (1, d, 1) ($d=0.5$ & $\phi=0.25$)					FIGARCH (1, d, 0) ($d=0.5$ & $\phi=0$)				
β	$T=100$	$T=200$	$T=500$	$T=1000$	β	$T=100$	$T=200$	$T=500$	$T=1000$
0.10	0.129	0.214	0.366	0.520	0.10	0.179	0.299	0.491	0.652
0.30	0.145	0.311	0.484	0.646	0.30	0.213	0.391	0.621	0.756
0.50	0.190	0.365	0.615	0.753	0.45	0.235	0.407	0.649	0.806
0.70	0.156	0.366	0.649	0.793	-	-	-	-	-
FIGARCH (1, d, 1) ($d=0.75$ & $\phi=0.05$)					FIGARCH (1, d, 0) ($d=0.75$ & $\phi=0$)				
β	$T=100$	$T=200$	$T=500$	$T=1000$	β	$T=100$	$T=200$	$T=500$	$T=1000$
0.10	0.129	0.188	0.298	0.349	0.10	0.151	0.192	0.331	0.349
0.30	0.193	0.293	0.436	0.498	0.30	0.192	0.286	0.443	0.533
0.50	0.275	0.422	0.616	0.700	0.50	0.294	0.441	0.657	0.744
0.70	0.326	0.563	0.773	0.890	0.70	0.319	0.525	0.784	0.914

Table 2. Power of the test.

	FIGARCH (1, d, 1) with $d = 0$ & $\phi = 0.975$				FIGARCH (0, d, 0) with $d = 0$			
Lambda	$T = 100$	$T = 200$	$T = 500$	$T = 1000$	$T = 100$	$T = 200$	$T = 500$	$T = 1000$
0.10	0.301	0.412	0.515	0.526	0.036	0.064	0.206	0.415
0.30	0.346	0.466	0.583	0.615	0.170	0.482	0.939	1.000
0.50	0.373	0.502	0.665	0.732	0.413	0.880	0.999	1.000
1.00	0.479	0.681	0.826	0.900	0.857	1.000	1.000	1.000
3.00	0.774	0.887	0.943	0.976	0.996	1.000	1.000	1.000
	FIGARCH (1, d, 1) with $d = 0.5$ & $\phi = 0.25$				FIGARCH (0, d, 0) with $d = 0.5$			
Lambda	$T = 100$	$T = 200$	$T = 500$	$T = 1000$	$T = 100$	$T = 200$	$T = 500$	$T = 1000$
0.10	0.194	0.400	0.631	0.798	0.171	0.274	0.462	0.600
0.30	0.290	0.497	0.759	0.854	0.161	0.278	0.517	0.625
0.50	0.467	0.713	0.859	0.919	0.209	0.397	0.607	0.745
1.00	0.773	0.921	0.984	0.998	0.374	0.593	0.783	0.875
3.00	0.984	0.999	1.000	1.000	0.693	0.895	0.976	0.989
	FIGARCH (1, d, 1) with $d = 0.75$ & $\phi = 0.05$				FIGARCH (0, d, 0) with $d = 0.75$			
Lambda	$T = 100$	$T = 200$	$T = 500$	$T = 1000$	$T = 100$	$T = 200$	$T = 500$	$T = 1000$
0.10	0.336	0.522	0.765	0.900	0.109	0.168	0.259	0.303
0.30	0.371	0.547	0.783	0.884	0.111	0.157	0.283	0.382
0.50	0.458	0.636	0.812	0.914	0.142	0.220	0.372	0.465
1.00	0.634	0.788	0.894	0.937	0.219	0.367	0.547	0.635
3.00	0.925	0.984	0.992	0.997	0.451	0.645	0.824	0.868

Table 4 reports the breaks detected using the AIT-ICSS algorithm for the given four series. We detect four break points in the S&P 500, three break points in FTSE 100, zero break points in Nikkei 225 and four break points in CAC 40 which represent the presence of (n + 1) distinct volatility regimes in the time series. It can be seen that most of the detected breaks cannot be related to major macroeconomic or political events and, hence, are probably spurious. This may well be due to the drawback associated with the AIT-ICSS algorithm that the test is severely oversized when long memory is present in the volatility of the series. Our findings are similar to what was found by Schwert [9] in that it is difficult to explain sudden changes in volatility with corresponding macroeconomic events.

5. CONCLUSION

Our simulation experiment indicates that the AIT-ICSS test is severely oversized and exhibits low power when long memory is present in the volatility of a time series. The empirical analysis also indicates that most of the breaks detected by the AIT-ICSS test cannot be related to major macroeconomic and political events and, hence, are probably spurious.

Table 3. Estimates of d in the FIGARCH model.

	S&P_500	FTSE_100	Nikkei_225	CAC_40
d	0.527#	0.512#	0.424#	0.588#
SE	(0.162)	(0.188)	(0.142)	(0.182)

and * mean significant at 1% and 5% level of significance. SE represents the standard error.

Table 4. Breaks detected.

Index (period)	Number of change points	Time period	Events
S&P_500	4	05-05-1982 to 03-08-1988	-
		10-08-1988 to 22-04-1992	-
		29-04-1992 to 15-07-1998	-
		22-07-1998 to 11-07-2007	Russian financial crisis
		18-07-2007 to 28-11-2012	Global financial crisis
FTSE_100	3	04-01-1984 to 08-07-1998	-
		15-07-1998 to 16-03-2003	-
		23-03-2003 to 17-10-2007	-
		24-10-2007 to 28-11-2012	Global financial crisis
Nikkei_225	0	-	No break detected
CAC_40	4	06-01-1988 to 05-08-1998	-
		12-08-1998 to 10-07-2002	Russian financial crisis
		17-07-2002 to 19-03-2003	-
		16-03-2003 to 09-01-2008	-
		16-01-2008 to 28-11-2012	Global financial crisis

NOTES

[1] We also perform simulation experiments by setting significance level at 1% as well as 10% and the results (not shown here and available on request) give rise to a similar inference as obtained in these simulations.

[2] We also perform the same analysis by incorporating breaks at 25th percentile as well as 75^{th}percentile and obtain similar results. (The results are not shown here and will be available on request).

REFERENCES

1. Poon, S.H. and Granger, C.W.J. (2003) Forecasting Volatility in Financial Markets: A Review. Journal of Economic Literature, 41, 478-539.

2. Ross, S.A. (1989) Information and Volatility: The No-Arbitrage Martingale Approach to Timing and Resolution Irrelevancy. Journal of Finance, 44, 1-7.

3. Inclán, C. and Tiao, G.C. (1994) Use of Cumulative Sums of Squares for Retrospective Detection of Changes of Variance. Journal of the American Statistical Association, 89, 913-923.

4. Sanso, A., Arago, V. and Carrion, J.L. (2004) Testing for Change in the Unconditional Variance of Financial Time Series. Revista de Economiá Financiera, 4, 32-53.

5. Malik, F., Ewing, B.T. and Payne, J.E. (2005) Measuring Volatility Persistence in the Presence of Sudden Changes in the Variance of Canadian Stock Returns. Canadian Journal of Economics, 38, 1037-1056.

6. Hammoudeh, S. and Li, H. (2008) Sudden Changes in Volatility in Emerging Markets: The Case of Gulf Arab Stock Markets. International Review of Financial Analysis, 17, 47-63.

7. Wang, P. and Moore, T. (2009) Sudden Changes in Volatility: The Case of Five Central European Stock Markets. Journal of International Financial Markets, Institutions and Money, 19, 33-46.

8. Newey, W.K. and West, K.D. (1994) Automatic Lag Selection in Covariance Matrix Estimation. Review of Economic Studies, 61, 631-654.

9. Schwert, G.W. (1989) Why Does Stock Market Volatility Change over Time? The Journal of Finance, 44, 1115-1153.

CHAPTER 7

SesIndexCreatoR: An R Package for Socioeconomic Indices Computation and Visualization

Benoît Lalloué[1,2], Séverine Deguen[2,3*], Jean-Marie Monnez[4], Cindy Padilla[2,3], Wahida Kihal[2,3], Denis Zmirou-Navier[2,3,5], Nolwenn Le Meur[2]

[1]In stitut Gustave Roussy, Paris, France

[2] EHESP Rennes, Sorbonne Paris Cité, Rennes, France

[3] Inserm, UMR IRSET, Institut de recherche sur la santé l'environnement et le travail - 1085, Rennes, France

[4] Lorraine University, CNRS, INRIA UMR 7502, Institut Elie Cartan, Lorraine, France

[5] Lorraine University, Medical School, Lorraine, France

ABSTRACT

In order to study social inequalities, indices can be used to summarize the multiple dimensions of the socioeconomic status. As a part of the Equit'Area Project, a public health program focused on social and environmental health inequalities; a statistical procedure to create (neighborhood) socioeconomic indices was developed. This procedure uses successive principal components analyses to select variables and create the index. In order to simplify the application of the procedure for non-specialists, the R package SesIndexCreatoR was created. It allows the creation of the index with all the possible options of the procedure, the classification of the resulting index in categories using several classical methods, the visualization of the results, and the generation of automatic reports.

KEYWORDS

Socioeconomic Status, Multidimensional Index, Principal Component Analysis, Hierarchical Classification, R

1. INTRODUCTION

When studying social inequalities, it is generally interesting to take into account the socioeconomic status (SES) of an individual, a neighborhood or a region rather than consider only one socioeconomic variable such as educational level or income. However, socioeconomic status is a complex and multidimensional concept which encompasses many aspects such as employment, income, education, housing and social bonds. All of these aspects can themselves be represented by various variables. To synthesize and consider these different aspects, one solution is to create a SES index.

There are already many existing SES indices, especially at the neighborhood level [1] -[9]. However, most of them use a small number of variables, combine variables with simple methods (such as Z-score) and/or select variables only from the litterature, which seems inappropriate for the purpose of the Equit'Area Project, a public health program focused on social and environmental health inequalities (http://www.equitarea.org), as detailed elsewhere [10]. Thus, a new statistical procedure to create neighborhood socioeconomic indices was developed. Basically, this procedure does not create an index from a set of determined and precise variables, but aims to select, from a large data set, variables which will compose the SES index. It is based on several successive principal component analyses and the whole procedure is detailed in the aforementioned article. It has already been successfully applied in several analyses aiming to study health or environmental inequalities [11] [12].

Compared to other existing approaches to compute indices, our procedure is a slightly more complex to understand and apply, especially for non-statisticians than for some other SES indices; we specifically developed our model in a R package [13], named SesIndexCreatoR. The package is freely available on the website of the Equit'Area project and on CRAN. The site http://www.equitarea.org/documents/packages_1.0-0/ contains the basic functions needed to run the procedure (in its entirety or only in some steps) and to obtain the corresponding SES index. The purpose of this package is to give tools as simple as possible to perform the procedure while keeping the various possibilities it offered, like using different data mining methods, adding illustrative units, or performing only one step of the procedure. Moreover, once the index is created, users can display all the results of the different analyses both in text and graphical output, and generate a report summary. The user may also create categories of this index with different methods (hierarchical clustering with or without k-nearest neighbors, quantiles, or intervals).

In this paper we present and illustrate the use of the SesIndexCreatoR package for Lille agglomeration (a large French metropolitan area). For further examples we recommend reading the works by Padilla et al. as mentioned above.

2. MATERIAL AND METHODS

2.1. Data

The example data provided in the SesIndexCreatoR package concerns one large city in France, Lille (Nord Pas de Calais region, northern France), and some adjacent municipalities. The statistical unit is the sub-municipal French census block groups (called IRIS) defined by the National Institute of Statistics and Economic Studies (INSEE). These units have an average of 2000 inhabitants and are constructed to be as homogeneous as possible in terms of socio-demographic characteristics and land use. Census block groups (BGs) are divided into three distinct categories: housing, economical activity and miscellaneous. Housing BGs are the most common, economical activity BGs include at least 1000 employees and at least twice as many employees as residents, and miscellaneous BGs are specific wide areas sparsely populated (leisure parks, port areas, forest, etc.). As activity and miscellaneous BGs have some particular profiles due to the way they are defined, they are treated in the example as illustrative units (meaning that they are not part of the procedure but will have an index value). For confidentiality and distribution reasons, the real BGs idenficators are replaced in the example data set with a simple number from 1 to 234 (which is the number of BGs of the area).

Socioeconomic data are taken from the 1999 national census (source: INSEE) and provide counts of population, households and residences at BG scale covering all the social, economic and demographic aspects. Median income at the BG scale is taken from a second database: the "Revenusfiscaux des ménages" database (source: INSEE-DGI). Using this raw data, 37 variables are defined at the BG scale based on the INSEE definitions. These variables are chosen to be representative of the theoretical concept of SES and in line with the variables most often used in the literature, or that could be considered as linked with the SES concept.

All variables are related to family structure, household type, immigration status, employment, income, education and housing (more details are available in Table 1 and Table 2). Some of the variables are intentionally redundant and represent the same notion, in view to determine which best represents this notion (using the algorithm implemented in the proposed package). In our example, there are two such groups: 7 variables of unemployment and 3 variables of labor force. We also note there are an unexpectedly high number of missing values for median income but, willing to keep this variable in the analysis, we filled missing values with the average value of the adjacent BGs.

2.2. SES Index Creation

The SES index creation procedure is detailed in Lalloué et al. [10]. Basically, it follows three successive steps:

(1) Study of the redundant variables. As already mentioned, several variables represent the same notion and we want to determine which best represented this notion. Therefore, one variable is selected for each group by applying principal component analysis (PCA) to each of the groups of redundant variables. The selected variable for each group is the one with the largest correlation with the first component of the PCA on the group.

(2) Selection of the variables. A PCA or a multiple factor analysis (MFA) on the remaining variables (i.e., non-redundant variables and variables selected in step (1) is used to select the variables with a contribution to the first component larger than the average one, i.e., variables that were best correlated with the first component ; i.e. variables that were best correlated with the first component. The choice of PCA or MFA depends on the willingness to give the same weight in the analysis to each domain (MFA) or not (PCA).

Table 1. Description of 37 socioeconomic variables available for the Lille agglomeration at the census block group scale, by domain. (Unless stated otherwise, variables are proportions expressed in %; a Redundant group “labor force”; b Redundant group “unemployment”; c Not a proportion).

Domain	*Variable name*	*Description*
BG type	Type	Census block group type (H: housing; A: activity; D: miscellaneous; Z: one BG municipality)[c]
Family and Household	UnderAge25	people under the age of 25 in the total population
	OverAge65	people over the age of 65 in the total population
	SingleParentFamilies	Single-Parent families in the total population
	HouseHolderAlone	Householders living alone in the total population
Immigration	ForeignForce	People in the labor force in the total population
Employment and Income	LabourForce	People in Labor Force in the total population[a]
	MenLabourForce	Men in labour Force in the male population[a]
	WomenLabourForce	Women in labour Force in the female population[a]
	UnemploymentTotal	Unemployment people in the labor force[b]
	UnemploymentForeigners	Unemployment Foreigners in the labor force[b]
	UnemploymentAge1524	Unemployment people in the 15 - 24 years old labor force[b]
	UnemploymentOverAge50	Over 50 years old unemployment people in the labor force[b]
	UnemploymentMen	Unemployment people in the male labor force[b]
	UnemploymentWomen	Unemployment people in the female labor force[b]
	UnemploymentMore year	People unemployed for more than 1 year in the labor force[b]
	SelfEmployed	Self-employed (independent workers, employers, etc.) in the labor force
	InsecureJobs	People with unstable jobs in the labor force
	SteadyJobs	People with steady jobs in the labor force
	MedianIncome	Median income per consumption unit (in euros per year)[c]

Table 2. Description of the 37 socioeconomic variables available for the Lille agglomeration at the census block group scale, by domain (continued). (Unless stated otherwise, variables are proportions expressed in %).

Domain	*Variable name*	*Description*
Education	AttendingSchool	People 6 - 15 years old school in the 6 - 15 years old population
	NoDiplomas	People with no diploma (and no studying) in the 15 years old and more population
	BasicGeneralQualifications	People with basic or intermediate general or vocation qualifications (and not studying) in the 15 years old and more population
	GeneralCertificates	People with general or vocational maturity certificates (and not studying) in the 15 years old and more population
	LowerTertiaryEducation	People with at least a lower tertiary education (and not studying) in the 15 years old and more population
	HigherEducationalDegree	People with a higher educational degree (and not studying) in the 15 years old and more population
	Students	Student in the 15 years old and more population
Housing	IndividualHouse	Individual houses in the main residences
	MultipleDweelingUnit	Multiple dwelling units in the main residences
	NonOwner	Non-Owner-occupied in the main residences
	SubsidizedHousing	Subsidized housing in the main residences
	BuildBefore1968	Main residences built before 1968
	BuildAfter1990	Main residences built after 1990
	Less40m^2	Main residences less than 40 m^2
	Larger150m^2	Main residences larger than 150 m^2
	ParkingSpace	Main residences with a parking space (garage or other)
	WithoutCar	Households without a car
	TwoOrMoreCars	Households with 2 or more cars

(3) Construction of the index. A final PCA is carried out including the variables selected in step (2) Provided that the first component of this PCA could be interpreted as a "SES component", it is used to calculate the socioeconomic index as the reduced first component.

3. THE SESINDEXCREATOR PACKAGE

The SesIndexCreatoR package depends on the FactoMineR [14] [15] and class [16]. In particular, most of data analysis and visualization functions, such as principal component analysis or hierarchical clustering, used in this package come from FactoMineR. We thus refer the user to the FactoMineR package and its manual for details on PCA and HC functions outputs. The sources and binaries of the package SesIndexCreatoR are available on the Equit'Area website or on CRAN and the installation is standard.

Because the package is also aimed to be used by R novice users, the example data are not included as R dataset but as a text file, in order to show in the package's manual how to import a file.

SesIndexCreatoR is composed of three main functions and several visualizing and internal functions (see Table 3):

Ÿ SesIndex function creates a socioeconomic index such as defined in the Equit'Area project. It is possible to choose the starting set of variables, the potential redundant groups of variables, the potential supplementary units, the method of selection (PCA or MFA) and the step of the procedure to perform. Results include the final index and all the results of the intermediate steps.

Ÿ SesClassif function creates socioeconomic categories, based on a socioeconomic index created by SesIndex function, with different technics such as hierarchical clustering, quantiles or equals subdivisions. Results include both a table with the original data set with class of each unit and the results of the classification technic (cut points, classes particularities,...).

Ÿ SesReport function creates a .html file with a report summarizing the results of the different steps of the creation of a socioeconomic index with the SesIndex function and, if any, the classification of the index using the SesClassif function. This function also allows to create a.csv file containing the original data set and the index and, if any, the classification.

4. EXAMPLE

First, the socioeconomic data from the text file are imported in a data frame:

```
R>library("SesIndexCreatoR")

R>SesData<- read.table(

+ system.file("extdata","SesData.txt", package = "SesIndex CreatoR"),

+ header=TRUE,sep="\t", row.names=1)
```

The SesData.txt contains 37 socioeconomic variables and 1 type variable (giving the type of BG) for each BG of the Lille municipality and adjacent municipalities, as describe in Section 2.1 Data. Then, the SesDatadataframe has 234 rows representing the BGs and 38 columns representing the variables.

Table 3. Functions available in SesIndexCreatoR 1.0-1.

Function	*Description*
ClassifHC	Internal function: Classification with Hierarchical clustering (HC)
ClassifInt	Internal function: Classification by intervals
ClassifQuant	Internal function: Classification by quantiles
plot.SesClassif	Plot the results of the classification of a socioeconomic index
plot.SesIndex	Plot the results of the construction of a socioeconomic index
print.SesClassif	Print the classification of a socioeconomic index results
print.SesIndex	Print the creation of a socioeconomic index results
SelectVar	Internal function: selection of variables
SesClassif	Create categories from a socioeconomic index
SesIndex	Creation of a Socio-Economic Index
SesReport	Creation of a report for SesIndex and SesClassif functions
SesStep1	Internal function: performs the first step of the creation of the socioeconomic index

As the SesIndex function needs vectors or lists of variables' names as arguments, we then extract the different vectors and lists needed to call the function (with redundant groups). The first line of the following code chunk allows to extract the names of the variables to analyse as a vector. The remaining lines extract the names of the variables in the two groups of redundant variables (seeTable 1) and create a list containing the two vectors of names for the groups of redundant variables.

```
R>varnames<- colnames(SesData)[2:ncol(SesData)]

R> group1 <- grep("+Unemployed", colnames(SesData), value=
TRUE)

R> group2 <- grep("+LabourForce", colnames(SesData), value=
TRUE)

R>groupvarnames<- list(group1, group2)
```

In order to consider activity and miscellaneous BGs as illustrative units, we extract the names of the corresponding rows (in our example, A is for "Activity" and D for "Miscellaneous" types of BGs):

```
R>illus<- rownames(SesData[SesData[,"Type"] %in% c("A",
"D"),])
```

It is "now" possible to create a socioeconomic index described in Materiel and methods using SesIndex. Here, we will create a socioeconomic index using all the 3 steps. Two groups of redundant variables are defined in groupvarnamesand several BGs are set illustrative. By default, all the 3 steps are performed and step 2 uses a PCA.

```
R> index <- SesIndex (SesData, varnames=varnames,
groupvarnames=groupvarnames,
```

```
+ sup=illus)

R>plot(index, choice="ind", label="none")
```

Once the index is created, we want to explore the results of the procedure. For instance, among the groups of redundant variables listed in Table 1 (Unemployment and Labor Force), the variables representing the best these groups and selected by our procedure are:

```
R> index$step1$selection

[1] "UnemployedTotal" "LabourForce"
```

Or, among the list of variables in Table 1 and Table 2 (except the redundant variables dropped at step 1), the variables selected to compose the SES index for Lille agglomeration are:

```
R> index$step2$selection

[1] "ForeignPop" "UnemployedTotal"

[3] "InsecureJobs" "SteadyJobs"

[5] "SingleParentFamilies" "NoDiplomas"

[7] "IndividualHouse" "MultipleDwellingUnits"

[9] "ParkingSpace" "NonOwner"

[11] "WithoutCar" "TwoOrMoreCars"

[13] "SubsidizedHousing" "MedianIncome"

R>plot(index, choice="var", step=2)
```

It is also possible to obtain detailed results of the data mining technics, like the correlation coefficients of the variables with the two first components of the second step analysis:

R> index$step2$analysisvarcoord[,c(1,2)]

	Dim.1	Dim.2
UnderAge25	0.63630110	0.21862821
OverAge65	−0.43163887	−0.18186596
ForeignPop	0.77678724	−0.15910609
LabourForce	−0.21536615	0.11435159
UnemployedTotal	0.87073008	−0.29624249
SelfEmployed	−0.54334383	0.47689134
InsecureJobs	0.89130598	0.13786739
SteadyJobs	−0.87777733	0.03703831
SingleParentFamilies	0.78556166	−0.19555464
NoDiplomas	0.64635948	−0.67461860
HouseholderAlone	0.45573846	0.72622501
AttendingSchool	−0.37025379	−0.02709806
BasicGeneralQualifications	−0.15542464	−0.87073704
GeneralCertificates	−0.62380039	0.12642215
LowerTertiaryEducation	−0.59037219	0.55626000
HigherEducationalDegree	−0.36942886	0.79823028
Students	0.35951016	0.68000017
IndividualHouse	−0.73553777	−0.46219873
MultipleDwellingUnits	0.74861629	0.43400686
BuiltBefore1968	0.09240263	−0.19191144
Builtafter1990	−0.05250108	0.60279988
ParkingSpace	−0.77646906	0.09453152
NonOwner	0.83998699	0.24679532
Less40 m^2	0.51827452	0.69853014
Larger150 m^2	−0.43976155	0.19532641
WithoutCar	0.90096827	0.17029037
TwoOrMoreCars	−0.87800029	−0.10183450
SubsidizedHousing	0.71268195	−0.27645939
MedianIncome	−0.82471346	0.17342006

Or the proportion of variance explained by the four first components of the final step:

R> index$step3$analysis$eig[1:4,]

```
eigenvalue percentage of variance
comp 1   9.4233115          67.309368
comp 2   1.6390996          11.707854
comp 3   1.0678887           7.627776
comp 4   0.5014364           3.581689
cumulative percentage of variance
comp 1                      67.30937
comp 2                      79.01722
comp 3                      86.64500
comp 4                      90.22669
```

The above outputs are especially interesting to understand the procedure of variable selection. We can see in these results that the variables of total unemployment and total labor force were respectively selected from the groups of redundant unemployment variables and labor force variables. Then, for these two groups only these two variables were kept in the next steps.

We can see in the selection from the step 2 that only variables with the highest correlations with the first component were selected. Here, 14 variables out of 29 were kept for the final step and the construction of the SES index.Eventually, the first component of the final step PCA performed on these 14 variables explained more than 67% of the total variance.

```
R>plot(index, choice="var", step=3)
```

Some graphical outputs can be seen in Figures 1-3. Figure 1 is a synthetic view of the projection of the BGs on the first principal components of the PCA performed in step 2 and step 3. Black dots represent active units whereas blue circles represent illustrative units (i.e., BG of the economical activity or miscellaneous types). Due to the number of units, BG labels are not displayed here but are activated by default. The step 3 part of the figure allows to see that BG are mainly along the first component and have not an extremely important variability along the second component.

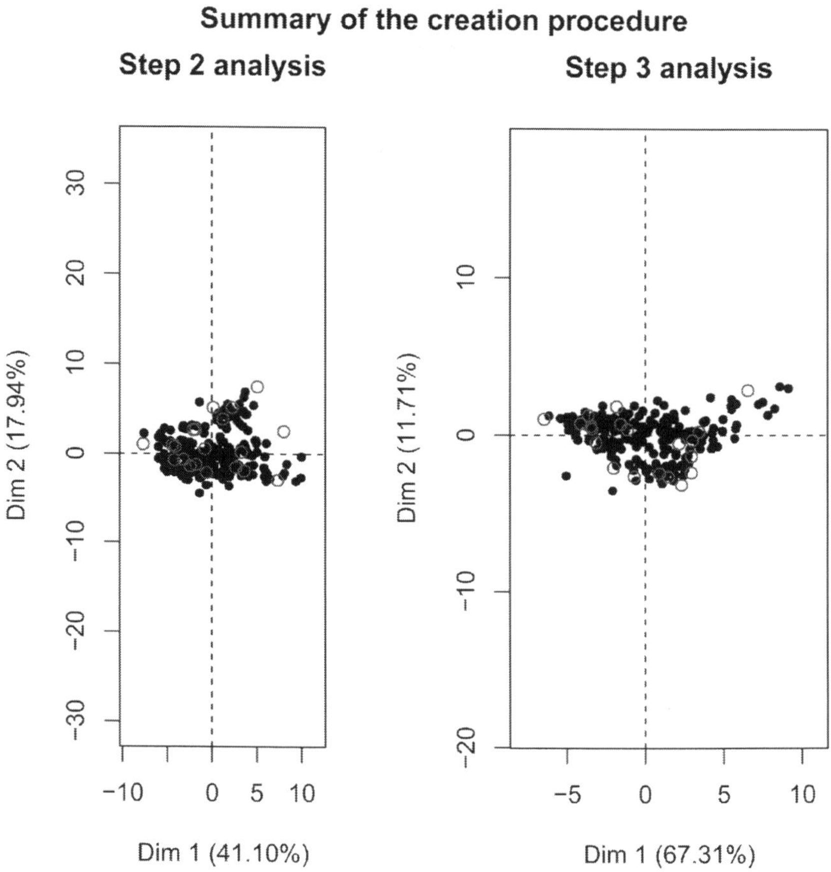

Figure 1. Synthetic view of the graphical outputs for individuals.

Figure 2 gives the circle of correlations of the PCA performed in step 2. Most of the variables seem to have a good correlation with the first component, both positively and negatively, whereas as the correlations with the second component are mainly positive (except for two variables). A few variables (5) are not well represented on this plane and may have higher correlations with the third or fourth component. On this figure, a first opposition between “variables of deprivation”, at the right, and “variables of favor”, at the left, can be seen.

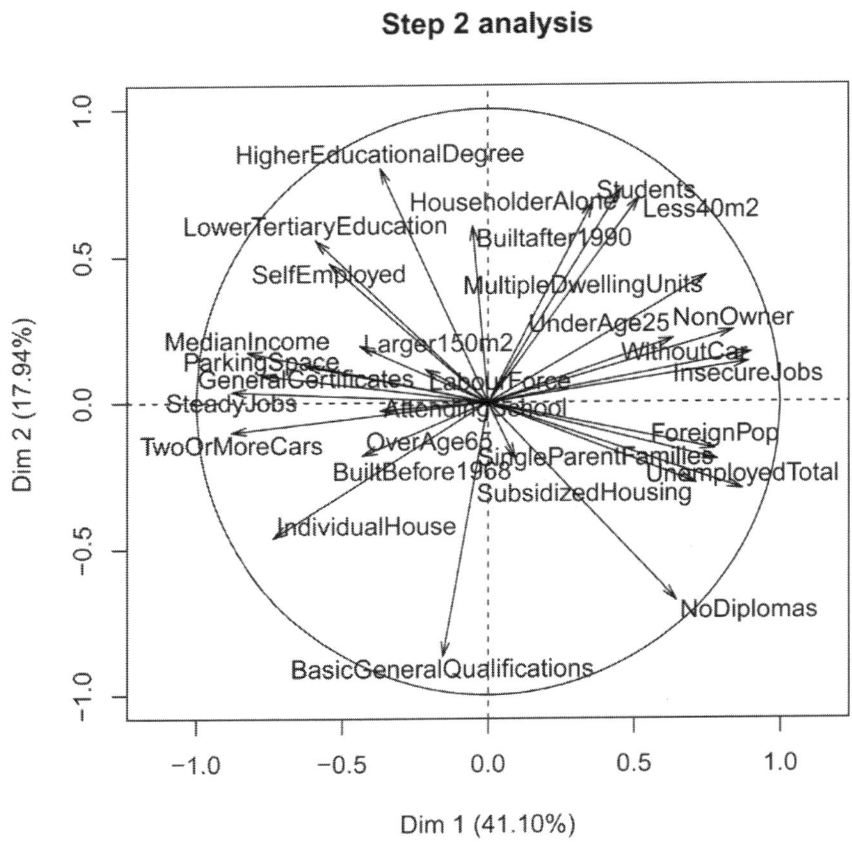

Figure 2. Correlation circle for the second step.

Finally, Figure 3 shows the circle of correlations of the PCA performed in step 3. The opposition between the "deprivation" and the "favor" variables is clear, with a high positive correlation between the first component and proportions of non-owner, unemployment, insecure jobs, person without diploma, subsidized housing, ... and a high negative correlation between the first component and proportion of steady jobs, individual houses, ... The first component of this PCA can then be interpreted as a SES component and be used as a SES index.

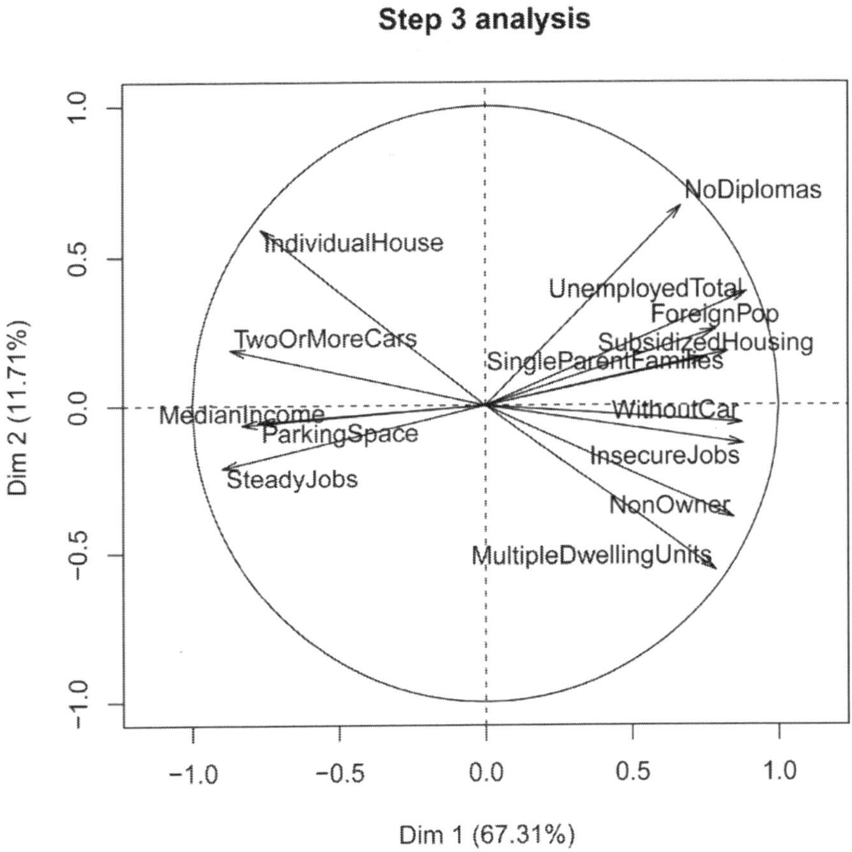

Figure 3. Correlation circle for the final step.

We now want to create categories from the socioeconomic index. We use a hierarchical clustering followed by a k-nearest neighbor (k-nn) algorithm. We decide to have an automatic number of classes (i.e., to cut the hierarchical clustering tree where the relative loss of inertia is the highest):

```
R> categories <- SesClassif(index)
```

Others possibilities currently in the SesClassif function are to create classes with hierarchical clustering without k-nn consolidation, with quantiles or with equal range of values.

We can summarize some characteristics of the different categories using simple functions. For instance, it is possible to compare variables average values in each category and the overall mean:

```
R> for (i in 1:3) {

+ print(paste("Category",i))

+
print(round(categories$analysis$desc.var$quanti[[i]][,c(2,3,6)],2))

+
```

[1] "Category 1"

	Mean in category	Overall mean	p.value
TwoOrMoreCars	0.33	0.21	0
IndividualHouse	0.71	0.45	0
SteadyJobs	0.73	0.65	0
ParkingSpace	0.60	0.43	0
MedianIncome	27529.06	21986.21	0
NoDiplomas	0.11	0.16	0
ForeignPop	0.02	0.05	0
SubsidizedHousing	0.08	0.26	0
UnemployedTotal	0.10	0.16	0
SingleParentFamilies	0.11	0.17	0
MultipleDwellingUnits	0.25	0.52	0
WithoutCar	0.16	0.29	0
InsecureJobs	0.09	0.13	0
NonOwner	0.35	0.58	0

[1] "Category 2"

	Mean in category	Overall mean	p.value
MultipleDwellingUnits	0.66	0.52	0.00
NonOwner	0.69	0.58	0.00
WithoutCar	0.35	0.29	0.00
InsecureJobs	0.14	0.13	0.00
SingleParentFamilies	0.18	0.17	0.02
SteadyJobs	0.63	0.65	0.01
MedianIncome	19693.52	21986.21	0.00
ParkingSpace	0.35	0.43	0.00
IndividualHouse	0.30	0.45	0.00

```
TwoOrMoreCars              0.14                  0.21        0.00
[1] "Category 3"
                           Mean in category Overall mean     p.value
UnemployedTotal            0.33                  0.16        0
ForeignPop                 0.14                  0.05        0
SingleParentFamilies       0.28                  0.17        0
SubsidizedHousing          0.74                  0.26        0
NoDiplomas                 0.30                  0.16        0
InsecureJobs               0.18                  0.13        0
WithoutCar                 0.47                  0.29        0
NonOwner                   0.90                  0.58        0
MultipleDwellingUnits      0.85                  0.52        0
IndividualHouse            0.13                  0.45        0
TwoOrMoreCars              0.08                  0.21        0
ParkingSpace               0.19                  0.43        0
MedianIncome               12624.39              21986.21    0
SteadyJobs                 0.46                  0.65        0
NULL
```

R>plot(categories$analysis, choice="map", label="none", draw.tree=F)

We can see that the optimal number of categories (according to the inertia criterion) was 3. The description of these categories showed that they are organised by decreasing socioeconomic status. Indeed, category 1 has higher average values of variables like median income or proportion of steady jobs, and lower average values of proportion of unemployed people or proportion of subsidized housing; whereas category 3 has lower values of median income and higher values of unemployment. Figure 4 shows the projection of these categories on the two first axes of the final PCA (note that it is also possible to use directly plot(categories) to have both the dendrogram and the projection of the units).

Eventually, we want to export the detailed results of all the three steps of creation of the SES index and of the classification. We also want to export a data file containing the index and the categories. To do so, the SesReport function is used to create .html report (see Appendix). By default, files are created in the current working directory with basename "SesReport" (which can be change as arguments of the SesReport function).

```
R>SesReport(categories)
```

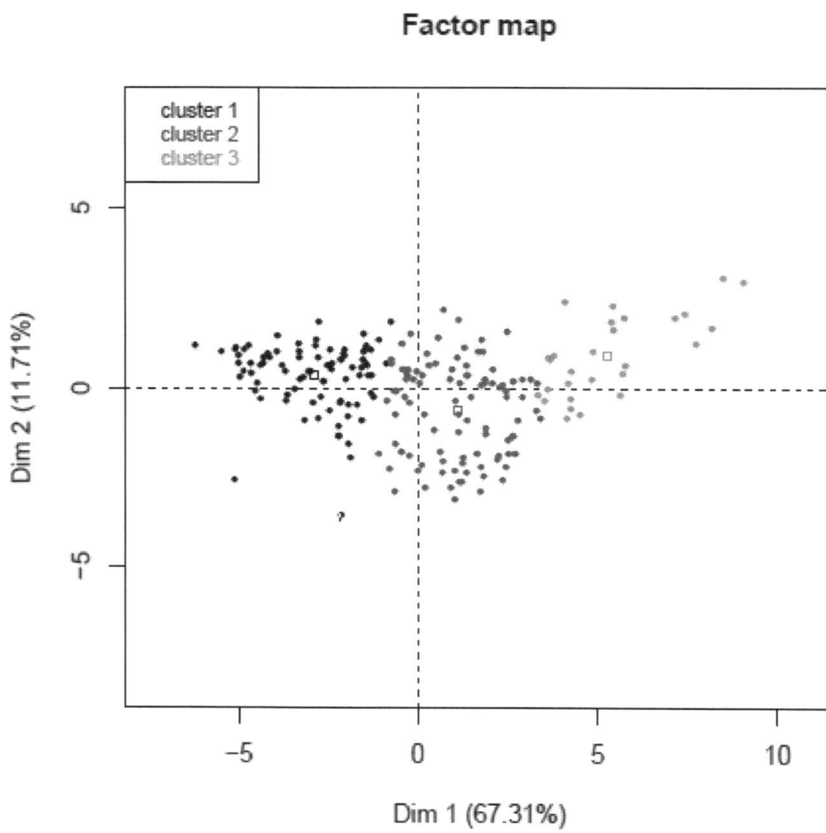

Figure 4. Plot of the individuals by categories.

5. CONCLUSIONS

In this article we presented the SesIndexCreatoR package, designed to easily create socio-economic indices with a reproductible statistical procedure. One originality of this procedure compared to other existing indices lays in selecting the final variables for the index by usage of data mining techniques rather than only information gleaned from a literature review, allowing to discard part of the subjectivity that may influence the choice of the variables. This data driven approach allows the data "speak by themself".

The SesIndexCreatoR package allow applying this procedure in a versatile way, by specifying which steps of the procedure should be runned (for instance only step 2 if the aim is to compare selection of variables between metropolitan areas without create indices, or only step 3 if one wants all the introduced variables to be in the index), adding illustrative units or selecting the method used. Once the index created, several tools are available to visualize, synthetize, explore and export the results in a convenient way for further utilization.

Benoît Lalloué, Séverine Deguen,, Jean-Marie Monnez, Cindy Padilla,, Wahida Kihal,, Denis Zmirou-Navier,,, Nolwenn Le Meur (such as mapping). However, these improvements will be made according to users' returns and needs.

REFERENCES

1. Jarman, B. (1983) Identification of Underprivileged Areas. British Medical Journal (Clinical Research Ed.), 286, 1705-1709.

2. Morris, R. and Carstairs, V. (1991) Which Deprivation? A Comparison of Selected Deprivation Indexes. Journal of Public Health, 13, 318-326.

3. Carstairs, V. (1995) Deprivation Indices: Their Interpretation and Use in Relation to Health. Journal of Epidemiology & Community Health, 49, S3-S8.

4. Salmond, C., Crampton, P. and Sutton, F. (1998) NZDep91: A New Zealand Index of Deprivation. Australian and New Zealand Journal of Public Health, 22, 835-837.

5. Eibner, C. and Sturm, R. (2006) US-Based Indices of Area-Level Deprivation: Results from Health Care for Communities. Social Science & Medicine, 62, 348-359.

6. Messer, L.C., Laraia, B.A., Kaufman, J.S., et al. (2006) The Development of a Standardized Neighborhood Deprivation Index. Journal of Urban Health, 83, 1041-1062.

7. Bell, N., Schuurman, N. and Hayes, M.V. (2007) Using GIS-Based Methods of Multicriteria Analysis to Construct Socio-Economic Deprivation Indices. International Journal of Health Geographics, 6, 17.

8. Fukuda, Y., Nakamura, K. and Takano, T. (2007) Higher Mortality in Areas of Lower Socioeconomic Position Measured by a Single Index of Deprivation in Japan. Public Health, 121, 163-173.

9. Pampalon, R., Hamel, D., Gamache, P., et al. (2009) A Deprivation Index for Health Planning in Canada. Chronic diseases in Canada, 29, 178-191.

10. Lalloué, B., Monnez, J.-M., Padilla, C., et al. (2013) A Statistical Procedure to Create a Neighborhood Socioeconomic Index for Health Inequalities Analysis. International Journal for Equity in Health, 12, 21.

11. Padilla, C., Lalloué, B., Pies, C., et al. (2013) An Ecological Study to Identify Census Blocks Supporting a Higher Burden of Disease: Infant Mortality in the Lille Metropolitan Area, France. Maternal and Child Health Journal, 17, 1-9.

12. Padilla, C.M., Deguen, S., Lalloue, B., et al. (2013) Cluster Analysis of Social and Environment Inequalities of Infant Mortality. A Spatial Study in Small Areas Revealed by Local Disease Mapping in France. Science of The Total Environment, 454-455, 433-441.

13. R Development Core Team (2011) R: A Language and Environment for Statistical Computing. R Foundation for Statistical Computing Vienna Austria Published Online First:

14. Husson, F., Josse, J., Le, S. and Mazet, J. (2013) FactoMineR: Multivariate Exploratory Data Analysis and Data Mining with R. R Package Version 1.25.

15. Lê, S., Josse, J. and Husson, F. (2008) FactoMineR: An R Package For multivariate Analysis. Journal of Statistical Software, 25, 1-18.

16. Venables, W.N. and Ripley, B.D. (2002) Modern Applied Statistics with S. 4th Edition. Springer, NY.

CHAPTER 8

Testing for Deterministic Components in Vector Seasonal Time Series

José Luis Gallego, Carlos Díaz

Departamento de Economía, Universidad de Cantabria, Santander, Spain

ABSTRACT

Certain locally optimal tests for deterministic components in vector time series have associated sampling distributions determined by a linear combination of Beta variates. Such distributions are nonstandard and must be tabulated by Monte Carlo simulation. In this paper, we provide closed form expressions for the mean and variance of several multivariate test statistics, moments that can be used to approximate unknown distributions. In particular, we find that the two-moment Inverse Gaussian approximation provides a simple and fast method to compute accurate quantiles and p-values in small and asymptotic samples. To illustrate the scope of this approximation we review some standard tests for deterministic trends and/or seasonal patterns in VARIMA and structural time series models.

KEYWORDS

Vector Time Series, Deterministic Components, Parametric Stability, Non-Invertibility, Unit Roots

1. INTRODUCTION

A wide class of test statistics for detecting the presence of deterministic components in univariate linear time series models can be derived following the [1] approach to the problem of testing for a scalar error covariance matrix in linear regression models with non-spherical disturbances. Members of this class are either locally best invariant (LBI) tests or LBI unbiased (LBIU) tests depending on whether the null hypothesis of a particular deterministic component is confronted with a one-sided or two-sided alternative, respectively. Some examples are the LBI test statistics for a null variance ratio or parametric stability proposed by [2-8], as well as the LBIU test statistics for non-invertibility or moving average (MA) unit roots derived by [9-12]. All these LBI and LBIU test statistics can be formulated as ratios of quadratic forms in normal variables, whose distribution functions are usually computed by numerical inversion of the corresponding characteristic functions using the [13] or [14] procedures. Moreover, their limiting distributions are related to that of the Cramèr-von Mises test statistics for goodness-of-fit derived and tabulated by [15]. Several non-parametric and parametric corrections have been proposed to cope with serially correlated errors so that the modified statistics follow the same limiting distributions.

Multivariate versions of these tests have been only derived in the framework of structural time series models by [5] and [16] based on the multivariate generalization of the [1] approach given by

[17]. As in the univariate case, the limiting distributions of the multivariate test statistics are also related to the Cramèr-von Mises distribution. However, the small sample distributions are unknown and must be evaluated by Monte Carlo simulation. A similar problem arises in the analysis of cointegrated VAR models with Dickey-Fuller type tests, where procedures for easily computing p-values and quantiles has been proposed among others by [18], using a response surface approach, and [19], fitting a Gamma distribution with moments estimated from a response surface regression. Approximating the unknown distribution of the tests for deterministic components has been also suggested by [17], who found that such distributions can be expressed as linear combinations of Beta variates and gave closed form expressions for their first two moments, which involve the computation of the eigenvalues of a matrix whose order depends on the sample size. However, a tentative family has not been proposed yet.

In this paper, we use the results of [17] to derive closed form expressions for the mean and variance of several test statistics for deterministic components that depends on the sample size and avoid the computation of eigenvalues. Besides, we propose a two-moment Inverse Gaussian (IG) approximation to the distribution of a linear combination of Beta variates. To illustrate some applications of this approximation we provide seasonal extensions of the [5] tests that can be used to test for non-invertibility in vector seasonal ARIMA models, as well as to derive easily the [16] tests for deterministic seasonality at specific frequencies.

The paper is organized as follows. In Section 2, we summarize the main results of the [17] approach. In Section 3, we review some relevant test statistics for deterministic components in vector time series and give exact expressions for the first two moments. In Section 4, we describe the two-moment IG approximation and assess its accuracy. Finally, in Section 5, we conclude with some extensions.

2. INVARIANT TESTS FOR COVARIANCE STRUCTURES

Consider the multivariate linear regression model

$$\mathbf{Y} = \mathbf{X}\Pi + \mathbf{E}, \; \mathbf{E} \sim N(0, \Omega(\delta) \otimes \Sigma) \tag{1}$$

where Y and E are T × m random matrices, X is a T × k fixed design matrix, P is a k × m matrix of parameters, W (d) and S are T × T and m × m positive definite matrices, respectively, and d is the parameter of interest determining whether or not the columns of E are i.i.d. Tdimensional errors, i.e., $W(d_0) = I_T$. [17] found that the Locally Best Invariant (LBI) test statistic of the null hypothesis H_0: d = d_0 against the one-sided alternative H_1: d > d_0 has the following general expression

$$L = \mathrm{tr}\left[\left(\hat{\mathbf{E}}'\hat{\mathbf{E}}\right)^{-1}\left(\hat{\mathbf{E}}'\mathbf{K}\hat{\mathbf{E}}\right)\right] \tag{2}$$

where tr is the trace operator, Ê = MY is the residual matrix in the ordinary least squares regression of Y on X, $M = I_T - X(X¢X)^{-1} X¢$ and K is the first derivative d W (d)/dd evaluated at d = d_0. Invariance is defined against the group of transformations YP + XA for an arbitrary k × m matrix A and a positive definite m × m matrix P. Thus, without loss of generality, it can be assumed that S = I_T. From [17], and following [1], it can also be proved that the LBIU test statistic of H_0: d = d_0 against the two-sided alternative H_1: d ≠ d_0 is given by (2) but with K being the second derivative $d^2W(d)/dd^2$ evaluated at d = d_0.

The null distribution of L can be characterized rewriting it as

$$L=\sum_{t=1}^{T\text{-}k}\lambda_t B_t,\quad B_t=\varepsilon_t'\left(\sum_{t=1}^{T-k}\varepsilon_t\varepsilon_t'\right)^{-1}\varepsilon_t, \tag{3}$$

where l_t are the non-null eigenvalues of the product matrix MK, e_t ~ $N(0, I_m)$, B_t ~ Beta $(m/2, (T - k - m)/2)$ and $B_1 + \ldots + B_{T-k} = m$, see, e.g., [20] p. 540. [17] found that the first two moments of L are given by

$E(L)=m\bar{\lambda}_1$ and $V(L)=c\left[\bar{\lambda}_2-\bar{\lambda}_1^2\right]$

with $c = 2m(T - k - m)/[(T - k - 1)(T - k + 2)]$ and

$$\bar{\lambda}_r=[1/(T-k)]\sum_{k=1}^{T-k}\mathrm{tr}\left[(\mathbf{MK})^r\right]$$

In the next section, we give formulae to compute tr (MK) and $tr(MK)^2$ for some useful test statistics.

3. TEST STATISTICS

3.1. Multivariate Seasonal Random Walk Plus Noise Model

[21] derived the LBI test statistic for a vector deterministic level in the multivariate local level model. We consider here the seasonal extension of this multivariate model given by

$$\mathbf{y}_t=\boldsymbol{\alpha}_t+\mathbf{u}_t,\ \boldsymbol{\alpha}_t=\boldsymbol{\alpha}_{t-k}+\mathbf{v}_t,\ t=1,\cdots,T, \tag{4}$$

where the vector time series $y_t = (y_{1t}, \ldots, y_{mt})$¢ is decomposed into the sum of a vector seasonal random walk $a_t = (a_{1t}, \ldots, a_{mt})$¢ plus a vector Gaussian white noise $u_t = (u_{1t}, \ldots, u_{mt})$¢ ~ N(0, S), the vector Gaussian white noise $v_t = (v_{1t}, \ldots, v_{mt})$¢ ~ N(0, rS) is assumed to be independent of u_t, and the parameter $r > 0$ quantifies the degree of stochasticity of a_t. Without loss of generality we assume that the seasonal period k is even and that the dataset is balanced, $T = nk$.

Defining the $T \times m$ matrices $Y = [y_1,\ldots, y_T]$¢, $A = [a_1,\ldots, a_T]$¢, $U = [u_1,\ldots, u_T]$¢ and $V = [v_1, \cdots , v_T]$¢, (4) can be written in matrix form as

$$\mathbf{Y} = \mathbf{A} + \mathbf{U}\,,\; (\mathbf{D}_n \otimes \mathbf{I}_k)\mathbf{A} = (\mathbf{d}_n \otimes \mathbf{I}_k)\mathbf{A}_0 + \mathbf{V} \tag{5}$$

where $A_0 = [a_{-k+1}, \ldots, a_0]$¢ is a $k \times m$ matrix of initial conditions, Ä denotes the Kronecker or tensor product, D_n is an $n \times n$ lower bidiagonal matrix with 1s on the main diagonal and –1s on the first sub-diagonal, which can be horizontally partitioned as $D_n = [d_n|$Ñ$_n]$¢, being $d_n = (1, 0, \ldots, 0)$¢ and Ñ$_n$ the $(n - 1) \times n$ first-order differencing matrix. If A_0 is assumed to be fixed, it follows that (4) is a special case of (1) with

$$\mathbf{X} = \mathbf{i}_n \otimes \mathbf{I}_k\,,\; \Pi = \mathbf{A}_0\,,\; \Omega(\rho) = \mathbf{I}_T + \rho(\mathbf{C}_n\mathbf{C}_n' \otimes \mathbf{I}_k)$$

where $\mathbf{C}_n = \mathbf{D}_n^{-1}$, $i_n = C_n d_n$ and X is a $T \times k$ matrix of seasonal dummy variables.

The LBI test statistic for testing the null hypothesis of deterministic seasonality (H_0: $r = 0$) against the alternative of seasonal random walk (H_1: $r > 0$), denoted by $RW_{m,k,n}$, is given by (2) with

$$\mathbf{K} = \mathbf{C}_n\mathbf{C}_n' \otimes \mathbf{I}_k \tag{6}$$

and Ê being the residual matrix in the multivariate regression of Y on the full set of k seasonal dummies. To compute the two first moments of RW_{mkn}, we find that the mean and mean-square of the eigenvalues of MK are given by

$\bar{\lambda}_1 = (n+1)/6$ and $\bar{\lambda}_2 = \bar{\lambda}_1(2n^2+7)/30$

which suggest to correct the $RW_{m,k,n}$ test statistic by a factor depending of the sample size so that it converges to a non-degenerate limiting distribution. Some candidates are n, mn or m (n – k). It should be noted that $RW_{m,k,n}/mn$ has asymptotic mean 1/6 and variance 1/45 mk.

[21] also derived the LBI test statistic for a vector deterministic linear trend. We obtain here the LBI test statistic for a vector deterministic seasonal linear trend by including a vector seasonal drift β_t in (4)

$$\mathbf{y}_t = \boldsymbol{\alpha}_t + \mathbf{u}_t,\ \boldsymbol{\alpha}_t = \boldsymbol{\alpha}_{t-k} + \boldsymbol{\beta}_t + \boldsymbol{v}_t,\ \boldsymbol{\beta}_t = \boldsymbol{\beta}_{t-k} \tag{7}$$

whose matrix form is given by

$$\mathbf{Y} = \mathbf{A} + \mathbf{U},$$

$$(\mathbf{D}_n \otimes \mathbf{I}_k)\mathbf{A} = (\mathbf{d}_n \otimes \mathbf{I}_k)\mathbf{A}_0 + (\mathbf{i}_n \otimes \mathbf{I}_k)\mathbf{B}_0 + \mathbf{V}$$

where $B_0 = [\beta_{-k+1}, \ldots, \beta_0]$¢ is a k × m fixed matrix of initial conditions for β_t If A_0 is fixed, (7) is a special case of (1) with X = [i_n Ä I_k, t_n Ä I_k], P = [A_0, B_0] and W(r) = I_T + r(C_n $\mathbf{C}'_n$ Ä I_k), where t_n = (1, 2, …, n)'. It is now clear that the inclusion of the vector seasonal drift only affects the mean vector of the sampling distribution of Y, $X\Pi$, but not its covariance matrix. Therefore, the LBI statistic for testing the null hypothesis of vector deterministic seasonal linear trend (H_0: r = 0) against vector seasonal drifted random walk (H_1: r > 0) in (7), say $DRW_{m,k,n}$, is computed as $RW_{m,k,n}$, being now Ê the residual matrix in the multivariate regression of Y on k seasonal dummies and k seasonal lineal trends. We find for $DRW_{m,k,n}$ that the mean and mean-square of the eigenvalues of MK are given by

$\bar{\lambda}_1 = (n+2)/15$ and $\bar{\lambda}_2 = \bar{\lambda}_1(11n^2+181)/840$

and so $DRW_{m,k,n}$ has asymptotic mean 1/15 and variance 11/6300 mk.

We also obtain another relevant modification of the $RW_{m,k,n}$ test statistic by including the time index t as a regressor in (4)

$$\mathbf{y}_t = \boldsymbol{\alpha}_t + \beta t + \mathbf{u}_t,\ \boldsymbol{\alpha}_t = \boldsymbol{\alpha}_{t-k} + \mathbf{v}_t \qquad (8)$$

or in matrix form,

$$\mathbf{Y} = \mathbf{A} + \beta \mathbf{t}_T + \mathbf{U},\ (\mathbf{D}_n \otimes \mathbf{I}_k)\mathbf{A} = (\mathbf{d}_n \otimes \mathbf{I}_k)\mathbf{A}_0 + \mathbf{V}$$

which is a special case of (1) with X = [i_n Ä I_k, t_T], P = [A_0, b_0] and W(r) as in (7). By the same token, the LBI statistic for testing H_0: r = 0 against H_1: r > 0 in (8), say $TRW_{m,k,n}$, is computed as $RW_{m,k,n}$, being now Ê the residual matrix in the multivariate regression of Y on k seasonal dummies and a regular lineal trend. We find for $TRW_{m,k,n}$ that the mean and mean-square of the eigenvalues of MK are given by

$$\bar{\lambda}_1 = [(5k-3)n^2 - (5k+3)]/\{30[(n-1)k-1]\}$$

and

$$\bar{\lambda}_2 = \frac{(140k-129)n^4 + (350k-213)n^2 - (490k+234)}{12600[(n-1)k-1]}$$

It should be noted that $DRW_{1,1,n} = TRW_{1,1,n}$ is the [3] test statistic for a univariate deterministic linear trend, and that the asymptotic mean and variance of $TRW_{1,1,n}/n$ are 1/15 and 11/6300, which agree with those obtained by [3] in a rather complicated proof.

3.2. Vector Seasonal IMA(1,1)$_k$ Model

Multivariate structural model (4) can be written as a vector seasonal IMA(1,1)$_k$ process

$$(\mathbf{I}_m - \mathbf{I}_m B^k)\mathbf{y}_t = (\mathbf{I}_m - \Theta \mathbf{I}_m B^k)\mathbf{a}_t, \quad \mathbf{a}_t \sim N(0, \Omega_a) \tag{9}$$

where B is the backshift operator such that $B^k y_t = y_{t-k}$, $a_t = (a_{1t}, \ldots, a_{mt})'$, W_a is a m × m positive definite matrix, and the parameter Θ is restricted to be positive so that $\rho = (1 - \Theta)^2/\Theta > 0$. Process (9) is said to be invertible when $\Theta < 1$ and strictly non-invertible when $\Theta = 1$. In the last case, the cancellation of the matrix polynomials on both sides of the equation reveals the presence of deterministic seasonality. Noting that $\rho(\Theta) = \rho(1/\Theta)$, the one-sided testing problem H_0: r = 0 versus H_1: r > 0 is equivalent to the two-sided one H_0: $\Theta = 1$ versus H_1: $\Theta \neq 1$. Hence, the LBI test statistic $RW_{m,k,n}$ for a null variance ratio in (4) is the LBIU test statistic for strict noninvertibility in (9). Note that $RW_{1,k,n}$ is the [11] test statistic for a seasonal MA unit root.

Analogously, it can be proved that $DRW_{m,k,n}$ and $TRW_{m,k,n}$ are the LBIU test statistics of H_0: $\Theta = 1$ versus H_1: $\Theta \neq 1$ in the reduced form of (7),

$$(\mathbf{I}_m - \mathbf{I}_m B^k)^2 \mathbf{y}_t = (\mathbf{I}_m - \mathbf{I}_m B^k)(\mathbf{I}_m - \Theta \mathbf{I}_m B^k)\mathbf{a}_t$$

and (8)

$$(\mathbf{I}_m - \mathbf{I}_m B)(\mathbf{I}_m - \mathbf{I}_m B^k)\mathbf{y}_t = (\mathbf{I}_m - \mathbf{I}_m B)(\mathbf{I}_m - \Theta \mathbf{I}_m B^k)\mathbf{a}_t$$

respectively. Note that $DWR_{1,k,n}$ is closely related to the [22] test statistic, while $TRW_{1,k,n}$ is the [12] test statistic for non-invertibility in the seasonal IMA(1,1)$_k$ model.

3.3. Dynamic Seasonal Linear Models

[6] considered testing the null hypothesis of deterministic seasonality against the alternative of mixed deterministic and stochastic seasonality. A related multivariate LBI test statistic can be derived in the seasonal linear regression

$$\mathbf{y}_t = \sum_{j=1}^{k} \mathbf{x}_{jt}\boldsymbol{\beta}_{jt} + \mathbf{u}_t, \quad \boldsymbol{\beta}_{jt} = \boldsymbol{\beta}_{jt-1} + \delta_j \mathbf{v}_{jt} \tag{10}$$

where x_{jt} ($j = 1, \ldots, k$) are a full set of seasonal dummy variables, β_{jt} is either a time-varying parameter if $d_j = 1$ or a nuisance constant parameter if $d_j = 0$, $u_t \sim N(0, S)$, $v_{jt} \sim N(0, (r/k)S)$ are mutually and serially uncorrelated vector errors, and is here divided by k for comparison purposes given that (10) reduces to (4) when $d_1 + \ldots + d_K = k$. Assuming that the initial conditions $\beta_{1,0}, \ldots, \beta_{k,0}$ are fixed, (10) is a special case of (1) with $X = [x_1, \ldots, x_k]'$, $P = [\beta_{1,0}, \ldots, \beta_{k,0}]'$ and $W(r') = I_T + r'(d_1A_1 + \ldots + d_kA_k)$, where $x_j = (x_{j1}, \ldots, x_{jT})'$, $A_j = x_j \circ C_T C'_T \circ x_j$, and the operator denotes the Hadamard product. The LBI test statistic for testing the null hypothesis of deterministic seasonality (H_0: $d_1 + \ldots + d_k = k$) against the alternative hypothesis of mixed deterministic-stochastic seasonality (H_1: $d_1 + \ldots + d_k = r < k$), say $SD_{m,k,n}$ (r), is given by (2) with $K = (d_1A_1 + \ldots + d_kA_k)/k$, which coincides with $RW_{m,k,n}$ when $r = k$. Noting that $MA_jMA_i = 0$ for $j \neq i$, we find that the eigenvalues of MK have mean and mean-square are given by

$$\bar{\lambda}_1 = (n+1)r/(6k) \text{ and } \bar{\lambda}_2 = \bar{\lambda}_1(2n^2+7)/30$$

Analogously, when the explanatory variables in (10) are trigonometric seasonal variables ($x_{1t} = 1$, $x_{jt} = \cos(jt\pi/k)$ for j even, and $x_{jt} = \sin[(j-1)t\pi/k]$ for j odd and $j > 1$), it is convenient to assume that $v_{jt} \sim N(0, r_jS)$, where $r_j = a_jr/k^2$ with $a_j = 1$ ($j = 1, k$) and $a_j = 2$ ($j = 2, \ldots, k-1$). Now, as before, (10) reduces to (4) when $r = k$. Here, we can focus our attention on testing the deterministic or stochastic nature of the local level β_{1t}, the (j/2)-th harmonic $\beta_{jt}\cos(\pi jt/k) + \beta_{j+1,t}\sin(\pi jt/k)$ ($j = 2, 4, \ldots, k$) or any combination of these k/2 harmonics. Taking as illustration the (j/2)-th harmonic, the LBI test statistic for testing the null hypothesis of deterministic seasonality (H_0: $d_1 + \ldots + d_k = k$) against the alternative of mixed deterministic-stochastic seasonality (H_1: $d_1 + \ldots + d_k = 2$) is given by (2) with $K = a_j (A_j + A_{j+1})/k^2$ with A_j as defined before. We find that the eigenvalues of MK have mean and mean-square given by

$$\bar{\lambda}_1 = a_j(n+1)/(6k)$$

and

$$\bar{\lambda}_2 = \bar{\lambda}_1\left[\frac{(4n^2-1)k^2+15}{60k^2} + b_j\frac{3}{4k^2\sin^2(j\pi/k)}\right]$$

where $b_j = 0$ ($j = 1, k$) and $b_j = 1$ ($j = 2, \ldots, k-1$). Note that these expressions are also valid to compute the mean and variance of the LBI test statistic for a deterministic level in presence of deterministic seasonality, H_0: $r_1 = 0$ versus H_1: $r_1 > 0$, which is closely related to the KPSS test with seasonal dummies proposed by [23]. Furthermore, [7,8] derived the LBI test statistic, say $TV_{1,k,n}$, for the dual testing problem of deterministic seasonality in presence of a deterministic level, which was generalized to the multivariate case by [16]. This test statistic is given by (2) with $K = (a_2A_2 + \ldots + a_kA_k)/k^2$. We find that the eigenvalues of MK have mean and mean-square given by

$$\bar{\lambda}_1 = \frac{(n+1)(k-1)}{6k} \quad \text{and} \quad \bar{\lambda}_2 = \bar{\lambda}_1\left[\frac{2n^2+7}{30} - \frac{k+1}{4k^2}\right]$$

4. APPROXIMATE DISTRIBUTIONS AND ACCURACY

4.1. Asymptotic Samples

Following [11] and [21], it can be shown that the limiting distribution of $RW_{m,k,n}/mn$ under testing H_0: r = 0 is

$$RW_{m,k,n} \rightarrow \frac{1}{mk}\sum_{j=1}^{\infty}(\pi j)^{-2}\zeta_j \tag{11}$$

where $\xi_j \sim$ iid χ^2_{mk}. [2] noted that $RW_{1,1,n}/n$ follows the same limiting distribution as the Cramèr-von Mises goodness-of-fit test statistic. Hence, (11) is the average of mk copies of the Cramèr-von Mises distribution, denoted by CvM(mk)/mk. [24] found that the first four cumulants of a CvM(r) distribution are given by $k_1 = r/6$, $k_2 = r/45$, $k_3 = 8r/945$, $k_4 = 8r/1575$ which reveal that the distribution is strongly rightskewed and leptokurtic. We observe that the two parameter Inverse Gaussian distribution, IG(μ, λ), can be fitted to possess similar characteristics. The first four cumulants of this distribution are $k_1 = \mu$, $k_2 = \mu^3/\lambda$, $k_3 = 3\mu^5/\lambda^2$, $k_4 = 15\mu^7/\lambda^3$ and matching the first two cumulants, μ = r/6 and μ^3/λ= r/45, we obtain that the fitted IG(r/6, $45r^2/6^3$) distribution has third and fourth cumulants given by κ_3 = 8r/900 and κ_4 = 8r/1350, which seem to be quite close to those of the CvM (r) distribution. The accuracy of the IG approximation is illustrated in **Figure 1**, which shows the limit pdf of $RW_{1,1,n}/n$ evaluated by the [13] procedure (solid line), along with the pdf of the fitted IG(μ, λ) distribution (dashed line) given by

$$f(x;\mu,\lambda) = [\lambda/(2\pi x^3)]^{1/2} \exp\,[-\lambda(x-\mu)^2/(2\mu^2 x)]$$

with x > 0, μ = 1/6 and λ = $45m/6^3$. We can see that the IG approximation provides a very good fit on both tails of the distribution. Therefore, accurate asymptotic pvalues for $RW_{m,k,n}/mn$ can be computed from the cdf of the IG(μ, λ) distribution.

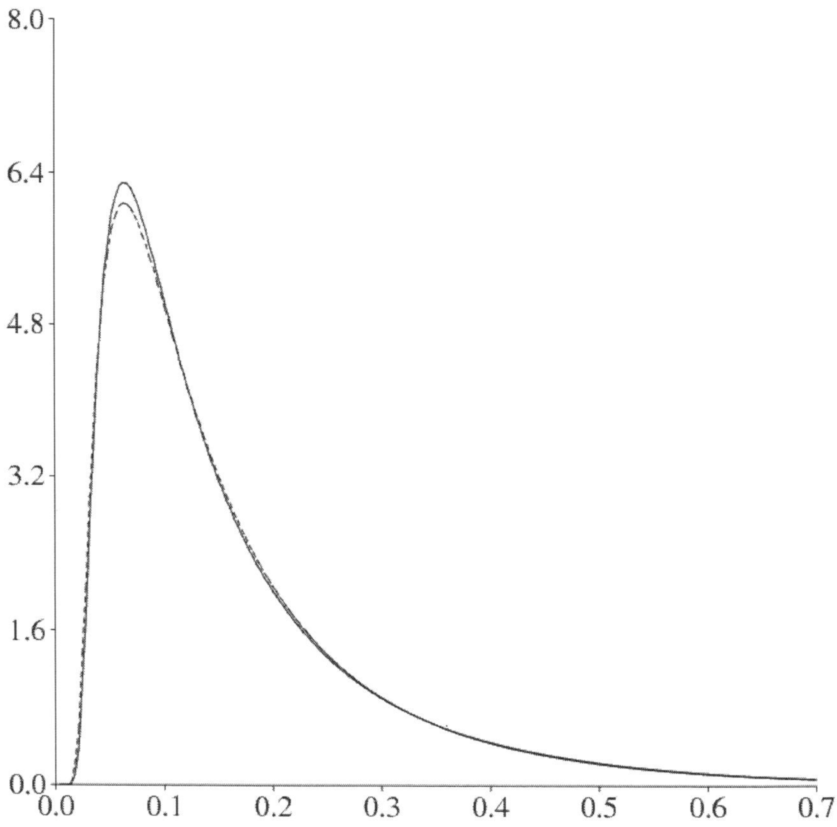

Figure 1. Accuracy of the IG approximation.

4.2. Finite Samples

The goodness-of-fit in the asymptotic case take us to ask if the approximation will be also good in finite samples. To evaluate the exact null distribution of $RW_{m,k,n}/mn$ from (3)-(6), we must determine the eigenvalues of the matrix MK. To this end, it is convenient to note that the projection matrix M can be alternatively written as M = [Ñ¢$_n$(Ñ$_n$Ñ¢$_n$)$^{-1}$Ñ$_n$] Ä I_k. Hence, MK = (Ñ$_n$Ñ¢$_n$)$^{-1}$ Ä $I_k$and its eigenvalues are the reciprocals of those of the tridiagonal matrix Ñ$_n$Ñ¢$_n$, $\lambda_t = [4\sin^2(t\pi/2n)]^{-1}$ (t = 1, 2, …, n – 1), each one with multiplicity k. In the case m = 1, (3)-(6) can be expressed as a ratio of quadratic forms in normal variables whose distribution was tabulated by [11] using the [13] procedure (they used n – 1 as correction factor instead of n). However, when m > 1, similar tables can be obtained by Monte Carlo simulation from (2)-(6).

To assess the accuracy of the IG approximation in small samples we simply compare the approximate pvalues with the nominal sizes for n = 10, 20, 30, 50 10, k = 1, 2, 3, and m = 1, 2, 3, 4, 5. In general, the approximate p-values agree closely with the nominal sizes even in small samples, being the mean absolute errors less than 0.003. Such discrepancies seem not to be relevant in practical applications. Similar results have been found for $DRW_{m,k,n}$ and $TV_{m,k,n}$. The results of this simulation study are not presented here due to space restrictions but are available from authors at request.

5. CONCLUSIONS

We have presented seasonal extensions of the [21] test for a deterministic level in multivariate models that can be used to detect different forms of non-invertibility in VARIMA models. The two-moment IG approximation to the null distribution of these test statistics, along with the closed forms expressions for the first two moments here derived, provide a simple and fast way to compute accurate critical values and p-values in practical applications. The proposed approximation could be also useful when modifying the test statistics to deal with intervention variables and serially correlated errors. Finally, the testing procedures described have been implemented in a computer program that can be freely obtained from authors at request.

REFERENCES

1. M. L. King and G. H. Hillier, "Locally Best Invariant Tests of the Error Covariance Matrix of the Linear Regression Model," Journal of the Royal Statistical Society, Series B (Methodological), Vol. 47, No. 1, 1985, pp. 98- 102.

2. J. Nyblom and T. Mäkeläinen, "Comparisons of Tests for the Presence of Random Walk Coefficients in a Simple Linear Model," Journal of the American Statistical Association, Vol. 78, No. 384, 1983, pp. 856-864.

3. J. Nyblom, "Testing for Deterministic Linear Trend in Time Series," Journal of the American Statistical Association, Vol. 81, No. 394, 1986, pp. 545-549.

4. D. Kwiatkowski, P. C. B. Phillips, P. Schmidt and Y. Shin, "Testing the Null Hypothesis of Stationarity against the Alternative of a Unit Root," Journal of Econometrics, Vol. 54, No. 1-3, 1992, pp. 159-178.

5. J. Nyblom and A. Harvey. "Testing against Smooth Stochastic Trends," Journal of Applied Econometrics, Vol. 16, 2001, pp. 415-429.

6. F. Canova and B. E. Hansen, "Are Seasonal Patterns Constant over Time? A Test for Seasonal Stability," Journal of Business and Economic Statistics, Vol. 13, No. 3, 1995, pp. 237-252.

7. M. Caner, "A Locally Optimal Seasonal Unit-Root Test," Journal of Business and Economic Statistics, Vol. 16, No. 3, 1998, pp. 349-356.

8. F. Busetti and A. Harvey, "Seasonality Tests," Journal of Business and Economic Statistics, Vol. 21, No. 3, 2003, pp. 420-436.

9. K. Tanaka, "Testing for a Moving Average Unit Root," Econometric Theory, Vol. 6, No. 4, 1990, pp. 433-444.

10. P. Saikkonen and R. Luukkonen, "Testing for a Moving Average Unit Root in Autoregressive Integrated Moving Average Models," Journal of the American Statistical Association, Vol. 88, No. 422, 1993, pp. 596-601.

11. W. Tam and G. C. Reinsel, "Tests for Seasonal Moving Average Unit Root in ARIMA Models," Journal of the American Statistical Association, Vol. 92, No. 438, 1997, pp. 725-738.

12. W. Tam and G. C. Reinsel, "Seasonal Moving-Average Unit Root tests in the Presence of a Linear Trend," Journal of Time Series Analysis, Vol. 19, No. 5, 1998, pp. 609-625.

13. J. P. Imhof, "Computing the Distribution of Quadratic Forms in Normal Variables," Biometrika, Vol. 48, No. 3-4, 1961, pp. 419-426.

14. R. B. Davies, "Numerical Inversion of a Characteristic Function," Biometrika, Vol. 60, No. 2, 1973, pp. 415-417.

15. T. W. Anderson and D. A. Darling, "Asymptotic Theory of Certain 'Goodness of Fit' Criteria Based on Stochastic Processes," The Annals of Mathematical Statistics, Vol. 23, No. 2, 1952, pp. 193-212.

16. F. Busetti, "Tests of Seasonal Integration and Cointegration in Multivariate Unobserved Component Models," Journal of Applied Econometrics, Vol. 21, 2006, pp. 419- 438.

17. J. Nyblom, "Invariant Tests for Covariance Structures in Multivariate Linear Model," Journal of Multivariate Analysis, Vol. 76, 2001, pp. 294-315.

18. J. MacKinnon, "Approximate Asymptotic Distribution Functions for Unit-Roots and Cointegration Tests," Journal of Business and Economic Statistics, Vol. 12, 1994, pp. 167-176.

19. J. A. Doornik, "Approximation to the Asymptotic Distributions of Cointegration Tests," Journal of Economic Surveys, Vol. 12, 1998, pp. 573-593.

20. C. R. Rao, "Linear Statistical Inference and its Applications," 2nd Edition, Wiley, New York, 1973.

21. J. Nyblom and A. Harvey, "Tests of Common Stochastic Trends," Econometric Theory, Vol. 16, 2000, pp. 176-199.

22. A. M. R. Taylor, "Locally Optimal Tests against Unit Roots in Seasonal Time Series Processes," Journal of Time Series Analysis, Vol. 24, No. 5, 2003, pp. 591-612.

23. P. C. B. Phillips and S. Jin, "The KPSS Test with Seasonal Dummies,"

24. B. M. Brown, "Cramèr-Von Mises Distributions and Permutation Tests," Biometrika, Vol. 69, No. 3, 1982, pp. 619-624.

CHAPTER 9

Unit Root Tests: The Role of the Univariate Models Implied by Multivariate Time Series

Nunzio Cappuccio [1] and Diego Lubian [2,*]

[1] Department of Economics and Management "Marco Fanno", via del Santo 33, Padova 35123, Italy

[2] Department of Economics, University of Verona, via dell'Artigliere 19, Verona 37129, Italy

ABSTRACT

In cointegration analysis, it is customary to test the hypothesis of unit roots separately for each single time series. In this note, we point out that this procedure may imply large size distortion of the unit root tests if the DGP is a VAR. It is well-known that univariate models implied by a VAR data generating process necessarily have a finite order MA component. This feature may explain why an MA component has often been found in univariate ARIMA models for economic time series. Thereby, it has important implications for unit root tests in univariate settings given the well-known size distortion of popular unit root test in the presence of a large negative coefficient in the MA component. In a small simulation experiment, considering several popular unit root tests and the ADF sieve bootstrap unit tests, we find that, besides the well known size distortion effect, there can be substantial differences in size distortion according to which univariate time series is tested for the presence of a unit root.

KEYWORDS

unit root tests; multivariate time series; cointegration

1. INTRODUCTION

It is well known that unit root tests may have large size distortion when the autoregressive parameter is close to unity and/or when there is a large MA component (see, for instance, [1]). The simulation evidence on the size distortion of some standard univariate unit root tests, such as the ADF test and Phillips-Perron $Z\alpha$ and Zt, is overwhelming (see, among others, [2]). Most simulation studies consider univariate unit root processes, but the same findings on the size distortion of unit root tests have been obtained by Reed [3,4] who considers a cointegrated VAR. He finds that the size distortion can be very large and not necessarily of similar magnitude across tests and univariate time series derived from the VAR. For instance, in the bivariate case, unit root tests applied to the one component may have an effective size as large as 90% while the same unit root test applied to the other component may have an effective size close to the nominal one. These finite sample results hold for DGPs characterized by quite different parameter values and for a wide range of roots of the AR component, and even in the case where the roots are 1 and 0.

In this note, we (a) provide a theoretical motivation for the finite sample size distortion observed in the presence of a large negative MA root; (b) give additional simulation evidence on its extent comparing standard and bootstrap unit root tests and (c) provide some suggestions for empirical researchers working with univariate time series implied by a VAR data generating process.

2. THE MODEL

We start the discussion from the VAR(1) model

$$\begin{pmatrix} y_t \\ x_t \end{pmatrix} = \begin{pmatrix} a_{11} & a_{12} \\ a_{21} & a_{22} \end{pmatrix} \begin{pmatrix} y_{t-1} \\ x_{t-1} \end{pmatrix} + \begin{pmatrix} u_{t1} \\ u_{t2} \end{pmatrix} \qquad (1)$$

or, in compact notation,

zt=Azt−1+ut,

where ut~i.i.d.(0,V). The representation for the univariate components, the so-called "final equations" in [5], can be obtained following [6]. Considering the lag polynomial matrix A(L)=I−AL, its determinant

$$|A(L)| = [(1 - a_{11}L)(1 - a_{22}L) - a_{12}a_{21}L^2]$$

and the adjoint matrix

$$A^*(L) = \begin{pmatrix} 1 - a_{22}L & a_{12}L \\ a_{21}L & 1 - a_{11}L \end{pmatrix},$$

the "final equations" for the VAR(1) model are given by

$$[(1 - a_{11}L)(1 - a_{22}L) - a_{12}a_{21}L^2] \begin{pmatrix} y_t \\ x_t \end{pmatrix} = \begin{pmatrix} 1 - a_{22}L & a_{12}L \\ a_{21} & 1 - a_{11}L \end{pmatrix} \begin{pmatrix} u_{t1} \\ u_{t2} \end{pmatrix}. \qquad (2)$$

It follows that the univariate processes evolve as an ARMA(2,1) model with a common AR component and two distinct MA components. 1

The magnitude and sign of the roots of the characteristic equation $|A-\lambda I|=0$ determine both the stationarity or nonstationarity of the univariate time series yt and xt and the existence of a cointegrating relationship between them. A necessary condition for cointegration is that the roots of the characteristic equation satisfy $\lambda 1=1$ and $|\lambda 2|<1$. From this unit root constraint, we obtain the restriction

$$a_{11} = 1 - \frac{a_{12}a_{21}}{1-a_{22}}, \tag{3}$$

which can be used to obtain the VECM representation

$$\Delta y_t = \alpha_1(y_{t-1} - \beta x_{t-1}) + u_{t1},$$
$$\Delta x_t = \alpha_2(y_{t-1} - \beta x_{t-1}) + u_{t1},$$

where $\alpha 1=-a12a21/(1-a22)$, $\alpha 2=a21$ and $\beta=(1-a22)/a21$.

The second restriction $|\lambda 2|<1$, *i.e.*,

$$-1 < \lambda_2 = a_{22} - \frac{a_{12}a_{21}}{1-a_{22}} < 1 \tag{4}$$

guarantees the stationarity of the error correction mechanism. In fact, it is easy to show that

$$(y_t - \beta x_t) = \lambda_2(y_{t-1} - \beta x_{t-1}) + (u_{t1} - \beta u_{t2}).$$

Imposing the constraints $\lambda 1=1$ and $|\lambda 2|<1$, the "final equations", *i.e.*, the univariate models become

$$(1-\lambda_2 L)\begin{pmatrix}\Delta y_t\\ \Delta x_t\end{pmatrix} = \begin{pmatrix}1-a_{22}L & a_{12}L\\ a_{21}L & 1-a_{11}L\end{pmatrix}\begin{pmatrix}u_{t1}\\ u_{t2}\end{pmatrix}. \tag{5}$$

Thus, if the DGP is a bivariate VAR(1) with one unit root and one cointegrating relationship, the marginal processes for the level processes follow an ARIMA(1,1,1) model and the marginal processes for the first-difference stationary processes are ARMA(1,1) processes. 2 It follows that the autocorrelation structure of the implied marginal processes, induced by the interaction of the AR and MA roots, is deemed to affect the finite sample size and power properties of unit root test in any simulation study where the DGP is a multivariate one.

Considering the right-hand side of (5), we see how the aggregate error term for each component of zt is the sum of an MA(1) process and a lagged white noise process

$$
\begin{aligned}
(1-\lambda_2 L)\Delta y_t &= \xi_{t1} = u_{t1} - a_{22}u_{t-1,1} + a_{12}u_{t-1,2},\\
(1-\lambda_2 L)\Delta x_t &= \xi_{t2} = u_{t2} - a_{11}u_{t-1,2} + a_{21}u_{t-1,1}.
\end{aligned}
$$

It is easy to show that both aggregate error terms on the right-hand side have the autocorrelation function of an MA(1) process so that we can write

$$
\begin{aligned}
\xi_{t1} &= v_{t1} + \theta_1 v_{t-1,1},\\
\xi_{t2} &= v_{t2} + \theta_2 v_{t-1,2},
\end{aligned}
\tag{6}
$$

where vt1 and vt2 are white noise processes. By setting the first-order autocorrelation coefficient of each marginal process on the left-hand side of (6) equal to the first-order autocorrelation coefficient of θi(L)vti, *i.e.*,

$$
\frac{\mathrm{Cov}(\xi_{ti}, \xi_{t-1,i})}{\mathrm{Var}(\xi_{t,i})} = \frac{\theta_i}{(1+\theta_i^2)},
$$

we can find the moving average coefficients of the polynomials θi(L) by choosing the invertible solution of the previous second degree equation. 3

For example, let us consider the DGP 4 in Table 1, *i.e.*,

$$A = \begin{pmatrix} 0.5 & 0.05 \\ 1 & 0.9 \end{pmatrix}, \quad V = \begin{pmatrix} 0.045 & 0.017 \\ 0.017 & 0.045 \end{pmatrix}.$$

The roots of the characteristic equation of the reduced form VAR are given by 1 and −0.4 and the marginal processes are given by

$$(1 - 0.4L)\Delta y_t = u_{t1} - 0.9u_{t-1,1} + 0.05u_{t-1,2},$$
$$(1 - 0.4L)\Delta x_t = u_{t2} - 0.5u_{t-1,2} + u_{t-1,1},$$

which, after some algebra, can be written as

$$(1 - 0.4L)\Delta y_t = v_t - 0.907v_{t-1},$$
$$(1 - 0.4L)\Delta x_t = v_t - 0.385v_{t-1}.$$

Thus, the MA component of the process Δyt has a large negative root, which explains the enormous size distortion, as reported in many simulation studies since Schwert [1].

Table 1. Parameter values in **MC(a)**.

DGP		β	ρ	θ_y	θ_x	$r_{\Delta y_t, \Delta x_t}$
DGP1						
$A = \begin{pmatrix} 0.1 & -0.1 \\ -0.9 & 0.9 \end{pmatrix}$,	$V = \begin{pmatrix} 0.061 & 0.086 \\ 0.0.86 & 0.32 \end{pmatrix}$	0.11	0	−0.71	0.12	0.258
DGP2						
$A = \begin{pmatrix} 0.5 & 0.1 \\ 0.5 & 0.9 \end{pmatrix}$,	$V = \begin{pmatrix} 1.47 & 1.41 \\ 1.41 & 2.32 \end{pmatrix}$	−0.2	0.4	−0.92	−0.69	0.821
DGP3						
$A = \begin{pmatrix} 0.7 & -0.9 \\ -0.3 & 0.1 \end{pmatrix}$,	$V = \begin{pmatrix} 0.045 & 0.017 \\ 0.017 & 0.045 \end{pmatrix}$	3	−0.2	0.14	−0.52	−0.195
DGP4						
$A = \begin{pmatrix} 0.5 & 0.05 \\ 1 & 0.9 \end{pmatrix}$,	$V = \begin{pmatrix} 0.045 & 0.017 \\ 0.017 & 0.045 \end{pmatrix}$	−0.1	0.4	−0.90	−0.38	0.592

Note: *A* and *V* are defined in (1).

3. A SIMULATION EXPERIMENT

In a small simulation study, we assess the size distortion of a number of unit root tests when the data comes from a cointegrated DGP. We consider the classical ADF test by Said and Dickey [10], the Zα and Zt tests proposed by Phillips and Perron [11], the modified MZα and MZt by Stock [12] and Perron and Ng [2], the modified Sargan-Bhargava MSB test proposed by Stock [12], the point optimal test PT of Elliott *et al.* [13] and its modification MPT proposed by Ng and Perron [14], and, finally the DF-GLS test by Elliott *et al.* [13]. We always estimate the spectral density at frequency zero of the error term using the autoregressive spectral density estimator as in Perron and Ng [2] and for the Zα,Zt,MZα,MZt,MSB tests we consider both OLS detrending and GLS detrending, as in Ng and Perron [14].

For the selection of the lag length, we do not follow a rule based just on the sample size but consider the Modified Akaike Information Criterion (MAIC) developed by Ng and Perron [14], where an upper bound to the lag length is set to the integer part of 12[(T+1)/100]1/4. Given the better performance of this information criterion compared to the BIC one, we do not consider the latter in our simulations. Further, we consider the suggestion by Perron and Qu [15] and present results for MZPQα obtained using OLS detrending instead of GLS detrending in the MAIC. They show that this simple modification produces tests with effective size closer to the nominal size.

We also consider two bootstrap unit-root tests. Palm *et al.* [16] carried out extensive simulation experiments on the size and power of bootstrap unit root considering ADF sieve and block bootstrap tests, based on first difference series or on residuals. Their findings suggest that, both in terms of size and power, the ADF sieve test as in Chang and Park [17] or its residual-based version perform best. In the following, we shall consider these two versions of the ADF sieve unit root test and implement the tests following the procedure set forth in Palm *et al.* [16].

Finally, to be able to make a comparison with a widely used test for the presence of unit roots and cointegration in multivariate time series, we also consider Johansen's trace-statistics, say Jtr, where, under the null, we should be able to reject no cointegration and not to reject the presence of one cointegrating vector.

In the simulation experiment, we consider two different ways of formulating the DGP. Firstly, as in Reed [3], we consider the VAR model in (1) subject to the constraints (3)–(4) which guarantee cointegration between yt and xt. Secondly, we also consider a DGP widely used in cointegration analysis (see, among others, [18]):

$$\begin{aligned} &y_t - \beta x_t = z_t, \qquad z_t = \rho z_{t-1} + e_{t1}, \\ &c_1 y_t - c_2 x_t = w_t, \quad w_t = w_{t-1} + e_{t2}, \\ &\begin{pmatrix} e_{t1} \\ e_{t2} \end{pmatrix} \sim \text{i.i.d.}N\left(\begin{pmatrix} 0 \\ 0 \end{pmatrix}\begin{pmatrix} 1 & \eta\sigma \\ \eta\sigma & \sigma^2 \end{pmatrix}\right). \end{aligned} \tag{7}$$

From this parameterization, we can obtain the implied VAR(1) as

$$\begin{aligned} \begin{pmatrix} y_t \\ x_t \end{pmatrix} &= \frac{1}{\beta_1 c_1 - c_2}\begin{pmatrix} \beta c_1 - c_2\rho & -c_2\beta(1-\rho) \\ c_1(1-\rho) & c_1\beta\rho - c_2 \end{pmatrix}\begin{pmatrix} y_{t-1} \\ x_{t-1} \end{pmatrix} + \frac{1}{\beta_1 c_1 - c_2}\begin{pmatrix} -c_2 & \beta \\ -c_1 & 1 \end{pmatrix}\begin{pmatrix} e_{t1} \\ e_{t2} \end{pmatrix} \\ &= \begin{pmatrix} a_{11} & a_{12} \\ a_{21} & a_{22} \end{pmatrix}\begin{pmatrix} y_{t-1} \\ x_{t-1} \end{pmatrix} + \begin{pmatrix} u_{t1} \\ u_{t2} \end{pmatrix}, \end{aligned}$$

where the unit root constraint a11=1−a12a21/(1−a22) is satisfied and under the condition

$$|\rho| = \left| a_{22} - \frac{a_{12}a_{21}}{(1-a_{22})} \right| < 1,$$

we have cointegration. The VECM representation is, then, given by

$$\begin{pmatrix} \Delta y_t \\ \Delta x_t \end{pmatrix} = \frac{1-\rho}{\beta_1 c_1 - c_2} \begin{pmatrix} c_2 \\ c_1 \end{pmatrix} (y_{t-1} - \beta x_{t-1}) + \begin{pmatrix} u_{t1} \\ u_{t2} \end{pmatrix} = \begin{pmatrix} \alpha_1 \\ \alpha_2 \end{pmatrix} (y_{t-1} - \beta x_{t-1}) + \begin{pmatrix} u_{t1} \\ u_{t2} \end{pmatrix}.$$

From the above VAR and VECM representations, we can obtain the DGPs considered in Reed [3] in terms of the parameters of (7) and *vice versa.*

In the simulation study, we consider DGPs parameterized, as in (7), following Gonzalo [18], and as in (1), following Reed [3]. The two set of parameters are defined as follows:

MC(a):

the values taken by A and V in Reed [3], reported in Table 1 together with the implied values of ρ, β, the roots of the MA component in the univariate representations and the unconditional contemporaneous autocorrelation;

MC(b):

as in Gonzalo [18], we set c2=−1, β=1 and c1=1 and consider the following values for the remaining parameters: ρ=(0.9,0.5), σ= (0.25,1,2) and η= (−0.5,0,0.5), for a total of 18 experiments. The root of the common autoregressive component (one root is always equal to 1) and the coefficients of the distinct MA components of the univariate models implied by the multivariate DGP are reported in Table 2.

All results are based on a sample size T=100, on 1000 simulations of the DGP and, for the bootstrap tests, on 500 bootstrap replications as in Psaradakis [19].

Table 2. Parameter values in **MC(b)**.

	AR root $\rho = 0.9$								
	$\sigma = 0.25$			$\sigma = 1$			$\sigma = 2$		
η	θ_y	θ_x	$r_{\Delta y_t, \Delta x_t}$	θ_y	θ_x	$r_{\Delta y_t, \Delta x_t}$	θ_y	θ_x	$r_{\Delta y_t, \Delta x_t}$
−0.5	−0.972	−0.978	0.911	−0.902	−0.942	0.029	−0.885	−0.924	−0.635
0	−0.975	−0.975	0.887	−0.930	−0.930	0.025	−0.910	−0.910	−0.583
0.5	−0.978	−0.972	0.911	−0.942	−0.902	0.029	−0.924	−0.885	−0.635
	AR root $\rho = 0.5$								
	$\sigma = 0.25$			$\sigma = 1$			$\sigma = 2$		
η	θ_y	θ_x	$r_{\Delta y_t, \Delta x_t}$	θ_y	θ_x	$r_{\Delta y_t, \Delta x_t}$	θ_y	θ_x	$r_{\Delta y_t, \Delta x_t}$
−0.5	−0.872	−0.893	0.925	−0.884	−0.884	0.910	−0.893	−0.872	0.925
0	−0.565	−0.717	0.158	−0.666	−0.666	0.142	−0.717	−0.565	0.158
0.5	−0.451	−0.626	−0.539	−0.565	−0.565	−0.500	−0.626	−0.451	−0.539

We notice that, for many values of the parameters, the univariate models are characterized by a large negative MA coefficient, which is exactly the circumstance in which unit root tests have low power and great size distortion even in moderately large sample sizes. For the set of parameters in**MC(b)**, this always occurs when ρ=0.9 and when ρ=0.5 and η=−0.5, while for the set of parameters in **MC(a)**, this occurs only in half of the cases.

Some additional remarks are in order. First of all, **MC(a)** is able to generate, at least for the parameter values considered here, greater heterogeneity in the MA roots of the univariate first-difference processes than that generated by **MC(b)**. In fact, in **MC(a)**, we observe a large negative MA root for Δyt associated to a small or a medium size negative MA root for Δxt or a positive MA root for Δyt associated to a negative one for Δxt. On the contrary, in **MC(b)**, looking, for instance, at the upper panel of Table 2 (the case in which ρ=0.9), we may see how the coefficients of the MA components are almost identical for the two processes Δyt and Δxt and that, to a lower extent, the same applies to the lower panel (ρ=0.5). When this occurs, the univariate representations of unit root processes resemble each other, and this may explain the similar behavior of unit root tests applied to yt and xt, separately.

Furthermore, when ρ=0.9, the coefficients of the MA component in Δyt and Δxt are not only very similar among themselves but also quite close to the root of the autoregressive component. This implies the presence of a near common factor in the univariate models for the first differenced series so that the AR and MA roots would almost cancel out. To our knowledge, this feature of the parameterization **MC(b)** used by (and many others Gonzalo [18]) has not been noticed so far in simulation studies on unit roots or cointegration tests. As a consequence of this near common root, the lagged unconditional correlation of Δyt and Δxt will tend to be small. For instance, for ρ=0.9, σ=2 and η=−0.5, the first-order unconditional autocorrelations of Δytand Δxt are equal to 0.015 and −0.021, respectively, and the first-order unconditional cross-correlation is equal to −0.032. Similar results are obtained when ρ=0.9 and for different values for σ and η. The first-order unconditional autocorrelations increase very slowly as ρ decreases, for instance, when ρ=0.5, we have 0.05, −0.11 and −0.16 for the first-order unconditional autocorrelation of Δyt, of Δxt and the first-order unconditional cross-correlation, respectively.

The empirical size, at a 5% nominal level, for the unit root tests, is reported in Table 3, Table 4,Table 5 and Table 6 for the estimated regression without a trend. 4 For each DGP, we test for a unit root both in yt and in xt. For each set of parameters in **MC(b)**, in each table, we report the effective size for a fixed value of the "signal-to-noise" ratio and different values of the remaining parameters; for the experiments in **MC(a)**, we report the effective size for the four different parameterizations.

The first general and striking result, common to both parameterization **MC(b)** and **MC(a)**, concerns the presence of important differences in the effective size according to whether yt or xtare tested for the presence of a unit root. Considering **MC(b)**, the empirical size increases with $\eta$when testing for a unit root in yt and, on the other hand, it decreases when testing is carried out on xt. This finding is remarkable and unexpected since the univariate ARMA representations of ytand xt share the same AR component and have very similar MA components for most parameterizations. The differences in size can be fully appreciated when σ=1 or σ=2. For instance, in the case σ=1,ρ=0.9,η=−0.5, for most tests based on GLS detrending (but for the MZPQt), the effective size is close to the nominal one when the unit root test is applied to yt, but it doubles or almost triples when unit root tests are applied to xt. In addition, the same applies to bootstrap unit root tests. The distortion in the effective size on yt and xt is reversed when σ=1,ρ=0.9 but η=0.5 and to a lesser extent for smaller values of the AR root such as ρ=0.5. Notice also that only in the case in which η=0, do these differences tend to to be negligible.

Table 3. Empirical size of unit root tests (no trend)—**MC(b)**, σ=0.25.

Test	$\rho = 0.9$						$\rho = 0.5$					
	$\eta = -0.5$		$\eta = 0$		$\eta = 0.5$		$\eta = -0.5$		$\eta = 0$		$\eta = 0.5$	
	y_t	x_t	y_t	x_t	y_t	x_t	y_t	x_t	y_t	x_t	y_t	x_t
OLS Detrending												
Z_α	0.029	0.058	0.037	0.030	0.044	0.027	0.018	0.054	0.049	0.048	0.065	0.023
MZ_α	0.013	0.025	0.014	0.014	0.018	0.018	0.009	0.029	0.021	0.027	0.032	0.009
Z_t	0.043	0.070	0.047	0.044	0.055	0.046	0.028	0.066	0.061	0.068	0.099	0.051
MZ_t	0.004	0.014	0.005	0.008	0.004	0.003	0.004	0.006	0.014	0.014	0.012	0.002
MSB	0.020	0.037	0.027	0.024	0.035	0.023	0.013	0.039	0.035	0.041	0.041	0.013
ADF	0.038	0.067	0.043	0.040	0.047	0.042	0.029	0.060	0.050	0.057	0.063	0.044
GLS detrending												
Z_α	0.047	0.084	0.063	0.048	0.088	0.046	0.040	0.077	0.070	0.070	0.098	0.043
MZ_α	0.041	0.071	0.052	0.041	0.077	0.039	0.034	0.07	0.059	0.062	0.083	0.036
MZ_α^{PQ}	0.005	0.009	0.008	0.004	0.007	0.008	0.006	0.012	0.010	0.013	0.008	0.004
Z_t	0.051	0.083	0.062	0.051	0.090	0.052	0.038	0.084	0.073	0.071	0.102	0.044
MZ_t	0.039	0.069	0.045	0.036	0.071	0.039	0.029	0.062	0.050	0.058	0.082	0.034
MSB	0.035	0.061	0.052	0.043	0.066	0.038	0.035	0.068	0.053	0.058	0.069	0.028
P_T	0.030	0.059	0.037	0.033	0.058	0.035	0.027	0.053	0.040	0.052	0.070	0.024
MP_T	0.037	0.067	0.042	0.035	0.071	0.037	0.029	0.061	0.049	0.059	0.082	0.032
$DF-GLS$	0.048	0.078	0.058	0.047	0.081	0.046	0.038	0.069	0.065	0.069	0.090	0.040
Bootstrap Tests												
ADF_b	0.045	0.058	0.060	0.043	0.064	0.036	0.040	0.051	0.048	0.053	0.062	0.043
ADF_b^r	0.044	0.062	0.062	0.056	0.070	0.040	0.042	0.061	0.059	0.057	0.064	0.044
Johansen's Trace Test												
J_{tr}	0.722		0.834		0.717		0.059		0.076		0.069	

Table 4. Empirical size of unit root tests (no trend)—**MC(b)**, σ=1.

	$\rho = 0.9$						$\rho = 0.5$					
	$\eta = -0.5$		$\eta = 0$		$\eta = 0.5$		$\eta = -0.5$		$\eta = 0$		$\eta = 0.5$	
Test	y_t	x_t	y_t	x_t	y_t	x_t	y_t	x_t	y_t	x_t	y_t	x_t
OLS detrending												
Z_α	0.033	0.098	0.082	0.080	0.107	0.037	0.049	0.146	0.108	0.119	0.164	0.081
MZ_α	0.014	0.057	0.045	0.040	0.062	0.020	0.026	0.090	0.060	0.071	0.087	0.038
Z_t	0.037	0.102	0.082	0.092	0.115	0.049	0.064	0.196	0.149	0.150	0.205	0.087
MZ_t	0.003	0.030	0.017	0.023	0.029	0.009	0.010	0.048	0.029	0.038	0.056	0.016
MSB	0.021	0.077	0.062	0.069	0.089	0.029	0.038	0.109	0.073	0.085	0.112	0.056
ADF	0.034	0.083	0.072	0.082	0.094	0.049	0.056	0.101	0.085	0.088	0.117	0.068
GLS detrending												
Z_α	0.064	0.156	0.137	0.122	0.184	0.069	0.065	0.155	0.126	0.156	0.184	0.100
MZ_α	0.053	0.134	0.118	0.102	0.154	0.056	0.055	0.115	0.100	0.122	0.145	0.078
MZ_α^{PQ}	0.004	0.030	0.019	0.018	0.028	0.007	0.013	0.038	0.031	0.033	0.043	0.016
Z_t	0.061	0.166	0.146	0.127	0.184	0.078	0.075	0.175	0.144	0.168	0.226	0.106
MZ_t	0.046	0.123	0.110	0.096	0.142	0.050	0.055	0.108	0.094	0.116	0.133	0.080
MSB	0.053	0.115	0.103	0.100	0.141	0.048	0.059	0.107	0.089	0.118	0.131	0.070
P_T	0.042	0.107	0.100	0.085	0.121	0.042	0.047	0.094	0.082	0.104	0.109	0.069
MP_T	0.047	0.118	0.111	0.095	0.143	0.049	0.055	0.107	0.093	0.115	0.132	0.076
$DF-GLS$	0.049	0.149	0.126	0.115	0.167	0.066	0.066	0.119	0.113	0.129	0.148	0.091
Bootstrap Tests												
ADF_b	0.049	0.130	0.105	0.091	0.125	0.052	0.067	0.096	0.077	0.079	0.099	0.061
ADF_b^r	0.052	0.141	0.110	0.107	0.136	0.055	0.071	0.099	0.083	0.090	0.109	0.072
Johansen's Trace Test												
J_{tr}	0.710		0.829		0.742		0.056		0.062		0.084	

Table 5. Empirical size of unit root tests (no trend)—**MC(b)**, σ=2.

	$\rho = 0.9$						$\rho = 0.5$					
	$\eta = -0.5$		$\eta = 0$		$\eta = 0.5$		$\eta = -0.5$		$\eta = 0$		$\eta = 0.5$	
Test	y_t	x_t	y_t	x_t	y_t	x_t	y_t	x_t	y_t	x_t	y_t	x_t
OLS Detrending												
Z_α	0.097	0.177	0.136	0.134	0.164	0.110	0.195	0.266	0.257	0.245	0.303	0.172
MZ_α	0.061	0.108	0.086	0.086	0.098	0.072	0.095	0.140	0.129	0.134	0.153	0.088
Z_t	0.083	0.159	0.121	0.116	0.153	0.103	0.254	0.423	0.364	0.347	0.432	0.247
MZ_t	0.028	0.053	0.040	0.034	0.052	0.037	0.052	0.101	0.096	0.090	0.106	0.050
MSB	0.079	0.139	0.116	0.115	0.134	0.100	0.120	0.162	0.152	0.160	0.176	0.11 0
ADF	0.072	0.141	0.108	0.100	0.124	0.089	0.125	0.158	0.148	0.149	0.161	0.112
GLS detrending												
Z_α	0.197	0.292	0.233	0.235	0.241	0.195	0.191	0.262	0.219	0.207	0.279	0.189
MZ_α	0.174	0.252	0.211	0.209	0.210	0.164	0.138	0.145	0.149	0.117	0.163	0.116
MZ_α^{PQ}	0.035	0.047	0.037	0.030	0.035	0.037	0.044	0.076	0.069	0.060	0.086	0.04
Z_t	0.209	0.297	0.246	0.235	0.256	0.214	0.240	0.381	0.307	0.279	0.369	0.238
MZ_t	0.157	0.234	0.192	0.197	0.194	0.164	0.134	0.130	0.138	0.116	0.155	0.106
MSB	0.157	0.237	0.187	0.200	0.205	0.150	0.118	0.130	0.138	0.109	0.150	0.109
P_T	0.137	0.200	0.171	0.172	0.170	0.138	0.119	0.112	0.122	0.099	0.132	0.089
MP_T	0.154	0.229	0.188	0.197	0.192	0.158	0.133	0.132	0.138	0.112	0.154	0.103
$DF-GLS$	0.188	0.271	0.223	0.221	0.235	0.199	0.149	0.168	0.156	0.126	0.176	0.122
Bootstrap Tests												
ADF_b	0.194	0.223	0.218	0.193	0.238	0.169	0.118	0.136	0.139	0.124	0.148	0.113
ADF_b^r	0.200	0.239	0.226	0.206	0.251	0.188	0.129	0.149	0.135	0.130	0.153	0.121
Johansen's Trace Test												
J_{tr}	0.730		0.849		0.711		0.070		0.052		0.059	

Table 6. Empirical size of unit root tests (no trend)—**MC(a)**.

	DGP1		DGP2		DGP3		DGP4	
Test	y_t	x_t	y_t	x_t	y_t	x_t	y_t	x_t
				OLS Detrending				
Z_α	0.927	0.086	0.283	0.016	0.140	0.739	0.672	0.041
MZ_α	0.084	0.024	0.112	0.009	0.017	0.020	0.188	0.011
Z_t	0.994	0.161	0.481	0.032	0.272	0.933	0.899	0.109
MZ_t	0.070	0.010	0.075	0.003	0.010	0.012	0.154	0.003
MSB	0.086	0.033	0.123	0.013	0.022	0.022	0.208	0.023
ADF	0.136	0.056	0.134	0.03	0.048	0.072	0.228	0.037
				GLS detrending				
Z_α	0.338	0.078	0.270	0.036	0.062	0.273	0.216	0.033
MZ_α	0.012	0.050	0.106	0.034	0.029	0.006	0.058	0.031
MZ_α^{PQ}	0.033	0.006	0.053	0.004	0.003	0.006	0.065	0.003
Z_t	0.559	0.094	0.393	0.034	0.096	0.490	0.343	0.029
MZ_t	0.012	0.047	0.101	0.032	0.028	0.006	0.057	0.028
MSB	0.010	0.046	0.093	0.031	0.032	0.005	0.052	0.031
P_T	0.012	0.035	0.085	0.029	0.027	0.006	0.054	0.024
MP_T	0.013	0.046	0.100	0.032	0.029	0.006	0.057	0.028
$DF-GLS$	0.062	0.059	0.133	0.028	0.051	0.052	0.093	0.039
				Bootstrap Tests				
ADF_b	0.136	0.052	0.104	0.039	0.061	0.085	0.189	0.044
ADF_b^r	0.122	0.051	0.113	0.034	0.056	0.078	0.188	0.041
				Johansen's Trace Test				
J_{tr}	0.044		0.064		0.041		0.037	

For the parameterization in **MC(a)**, in Table 6, we continue to observe substantial differences in the effective size according to whether yt or xt are being tested for a unit root. These differences are even more pronounced than those in Table 3, Table 4 and Table 5, perhaps because of the greater range and heterogeneity of the MA component obtained from the parameter values under**MC(a)**. Furthermore, there are noticeable differences among tests and across DGPs: for instance, looking at DGP1, both the DF-GLS and bootstrap tests have reasonable effective sizes for xt while the effective size more than doubles for the bootstrap tests applied to yt, but it does not change substantially for the DF-GLS test. Again, for DGP4, the size of the DF-GLS more than doubles when yt is tested while the size of the bootstrap tests is more than four times larger for yt than for xt.

Considering the parameterization in **MC(b)**, from Table 3, Table 4 and Table 5, we notice that when σ=0.25, the empirical size is, in general, close to the nominal one for most tests and it is so, in particular, for the ADF sieve bootstrap unit root tests. In particular, the empirical size in both versions of the ADF boostrap test seem to be more stable and closer to the nominal size than the empirical size of the PT, MPT, and DF-GLS tests. However, for σ=1, and to a larger extent when σ=2, the empirical size of these tests tend to differ more and more from the nominal 5% level. In fact, the effective size increases with σ ranging in the interval (0.03,0.09) when σ=0.25 to the interval (0.11,0.25) when σ=2. Thus, as the variance of the random walk component in yt and xt increases, the size of the unit root tests increases, leading to greater size distortion, and the size distortion itself is quite sizeable for σ=2, irrespective of ρ and η. In general, GLS detrending tends to increases the empirical size and this exerts a beneficial effect when σ is small, but, on the contrary, it is detrimental for the size when σ is large.

Considering the four parameterizations under **MC(a)**, the behavior of unit root tests is more heterogeneous, as is the pattern of AR and MA roots. Under OLS detrending, the Zα and Zt tests have large size distortion for all parameterizations. The modifications suggested by Perron and Ng [2] are somehow effective in reducing the distortion, but the behavior of the *M* tests is not stable across DGPs, and the same remark applies to the ADF test. Under GLS detrending, the DF-GLS test by Elliott *et al.* [13] has the best performance, showing an effective size very close to the nominal one in all cases but for yt in DGP2 and DGP4 where, in fact, the MA component has a coefficient close to −1. The bootstrap unit root tests have greater size distortion than the DF-GLS but, for yt in DGP2 and DGP4, exactly in those cases where the DF-GLS test does not perform well.

Finally, we consider consider Johansen's trace test under the null of one cointegrating vector. For parameters in **MC(b)**, the test statistics are severely biased when ρ=0.9, irrespective of the values taken by σ and η, while it has a size close to the nominal one for ρ=0.05 and its performance, in the latter case, is superior to those of the standard and bootstrap unit root tests. For the parameterizations in **MC(a)**, Johansen's trace test also displays a very good behavior since about 5% of the time, we reject no cointegration in favor of stationarity of the VAR in (1) in all cases. However, we should bear in mind that the AR root is rather small now and that, from the previous results based on **MC(b)**, Johansen's test is adversely affected by large values of ρ.

We do not consider the power properties of the unit root tests considered here. Ng and Perron [14] provide simulation evidence that the DF-GLS has better power then the *M* tests, even though the latter have better size properties. For the bootstrap unit root tests, we take the results of the extensive simulation study by Palm *et al.* [16] who found that the ADF sieve bootstrap test performs better under a variety of DGPs with and without an MA component. An extensive simulation study on the power properties of the tests considered here for univariate time series generated by a cointegrated VAR is left for a further investigation.

4. CONCLUSIONS

A standard practice in cointegration analysis is to run unit root tests separately for each single time series in the multivariate system. However, univariate time series are most often observed as a part of a more general multivariate model. It turns out that the univariate models implied by a VAR data generating process always have a finite order MA component (e.g., see [6]). This feature may explain why an MA component has often been found in univariate ARMA models for economic time series and, given the well-known size distortion of popular unit root tests in the presence of a large negative coefficient in the MA component, it has important implications for unit root tests in univariate settings. In a small simulation experiment under cointegration, we find that (a) there can be substantial differences in the size distortion according to whether the unit root test is applied on yt or xt and that this occurs for the ADF sieve bootstrap unit root test too; (b) most tests perform well when the "signal-to-noise" ratio is small but the size distortion can be large when the "signal-to-noise" ratio increases; (c) the ADF sieve bootstrap unit root tests are not immune from size distortion; and (d) Johansen's trace test based on the VAR model exhibits great size distortion when the root of the AR component is large.

ACKNOWLEDGMENTS

We would like to thank three anonymous referees and the editor for very helpful comments.

AUTHOR CONTRIBUTIONS

Both authors contributed equally to the paper.

NOTES

[1] In general, considering a *k*-dimensional VAR(p) process the univariate models will be at most ARMA(kp,(k−1)p), all univariate processes share the same AR component, and an MA component is present in each univariate model. See [5] for a general treatment of this issue.

[2] See Cubadda *et al.* [7] for a general result on the implied univariate models from cointegrated VAR and Cubadda and Triacca [8] for the I(2) case.

[3] A general solution to this problem for an MA process of order *q* has been provided in Maravall and Mathis [9].

[4] For brevity, our discussion will refer to the model without trend, similar remarks apply when a trend component is included in the regression, simulation results in this case are available upon request from the authors.

REFERENCES

1. Schwert, W. Testing for unit roots: a montecarlo investigation. *J. Bus. Econ. Stat.* **1989**, *7*, 147–159.

2. Perron, P.; Ng, S. Useful modifications to some unit root tests with dependent errors and their local asymptotic properties. *Rev. Econ. Stud.* **1996**, *63*, 435–563.

3. Reed, W.R. *Unit Root Tests, Size Distortion, and Cointegrated Data*; Working Paper; University of Canterbury: Christchurch, New Zealand, 2014.

4. Reed, W.R. *Univariate Unit root Tests Perform Poorly When Data are Cointegrated*; Working Paper; University of Canterbury: Christchurch, New Zealand, 2016.

5. Zellner, A.; Palm, F. Time series analysis and simultaneous equation econometric models. *J. Econom.* **1974**, *2*, 17–54.

6. Tsay, R.S. *Analysis of Financial Time Series*, 3rd ed.; John Wiley & Sons: New York, NY, USA, 2010.

7. Cubadda, G.; Hecq, A.; Palm, F. Studying co-movements in large multivariate data prior to multivariate modelling. *J. Econom.* **2009**, *148*, 25–35.

8. Cubadda, G.; Triacca, U. An alternative solution to the autoregressive paradox in time series analysis. *Econ. Model.* **2011**, *28*, 1451–1454.

9. Maravall, A.; Mathis, A. Encompassing univariate models in multivariate time series. *J. Econom.* **1994**, *61*, 197–233.

10. Said, S.E.; Dickey, D.A. Testing for unit roots in autoregressive-moving average models of unknown order. *Biometrika* **1984**, *71*, 599–607.

11. Phillips, P.C.B.; Perron, P. Testing for a unit root in time series regression. *Biometrika* **1987**,*75*, 335–346.

12. Stock, J.H. A Class of Tests for Integration and Cointegration. In *Cointegration, Causality and Forecasting: A Festschrift for Clive W.J. Granger*; Oxford University Press: Oxford, UK, 1999; pp. 135–167.

13. Elliott, G.; Rothenberg, T.J.; Stock, J.H. Efficient tests for an autoregressive unit root.*Econometrica* **1996**, *64*, 813–836.

14. Ng, S.; Perron, P. Lag length selection and the construction of unit root tests with good size and power. *Econometrica* **2001**, *69*, 1519–1554.

15. Perron, P.; Qu, Z. A simple modification to improve the finite sample properties of ng and perron's unit root tests. *Econ. Lett.* **2007**, *94*, 12–19.

16. Palm, F.C.; Smeekes, S.; Urbain, J. Bootstrap unit root tests: Comparisons and extensions. *J. Time Ser. Anal.* **2008**, *29*, 371–401.

17. Chang, Y.; Park, J.Y. A sieve bootstrap for the test of a unit root. *J. Time Ser. Anal.* **2003**, *24*, 379–400.

18. Gonzalo, J. Five alternative methods of estimating long-run equilibrium relationships. *J. Econom.* **1994**, *60*, 203–233.

19. Psaradakis, Z. Bootstrap tests for an autoregresive unit root in the presence of weakly dependent errors. *J. Time Ser. Anal.* **2001**, *22*, 577–594.

CHAPTER 10

Capital Asset Pricing Model Testing at Warsaw Stock Exchange: Are Family Businesses the Remedy for Economic Recessions?

Jacek Lipiec

Collegium of Business Administration, Warsaw School of Economics, 02-554 Warsaw, Poland

ABSTRACT

In this article, we test the capital asset pricing model (CAPM) on the Warsaw Stock Exchange (WSE) by measuring the performance of two portfolios composed of construction firms: family-controlled and nonfamily controlled. These portfolios were selected from the WIG-Construction (WIG—Warszawski Indeks Giełdowy—Warsaw Stock Exchange Index). The performance of both portfolios was measured in the period from 2006 to 2012 with respect to three sub-periods: (1) pre-crisis period: 2006–2007; (2) crisis period: 2008–2009; and (3) post-crisis period: 2010–2012. This division was constructed in this way to find out how family firms performed in crisis times in relation to nonfamily firms. In addition, the construction portfolio was chosen due to its sensitivity to recessions. When an economy faces a downturn, construction firms are among the first to be exposed to risk. The performance was measured by using the capital asset pricing model with statistical inference. We find that public family firms significantly outperformed non-family peers in the crisis times.

KEYWORDS

CAPM; beta; portfolio return; Polish family firms; Warsaw Stock Exchange

1. INTRODUCTION

The most recent U.S. domestic housing subprime crisis in 2007–2008 predominantly impacted the construction sector. Its negative consequences spread out to the rest of the U.S. economy and eventually expanded worldwide (International Labour Organization [1]). According to The Organisation for Economic Co-operation and Development (OECD) data, almost all countries tracked by the OECD reported a fall in GDP of 2.1 percentage points in the first quarter of 2009 and a rise in unemployment of 2.4 percentage points in May, 2009 (OECD [2]). This was the largest decline reported by the OECD since 1960, which predominantly affected the construction and automobile sectors. In the U.S., this impact was more devastating and led to an almost tripling of the unemployment rate, which reached 20.6% in 2010 (Current Population Survey (CPS) [3]). In response, the U.S. launched the American Recovery and Reinvestment Act of 2009. This crisis also had a very deep and negative impact on EU countries and the Euro area, where the construction sector accounted for 6.0% of EU GDP in 2010. The deepest decline of −8.5% in the EU-27 and −7.9% in the Euro area hit in 2009 (see Table 1).

Only one EU country reported positive growth in the construction sector during the crisis period: Poland with a growth of 10.2% in 2008 and 4.5% in the following year (see Table 2).

Due to its market vulnerability, the construction sector may be used as a litmus test for company performance and in a particular crises resistance.

It seems interesting to measure the behavior of family firms exposed to crises. Some researchers argue that during periods of economic meltdown, family businesses exposed to increased risk situations may perform differently than non-family business by prioritizing the security of family over firm interests (Lee [4]; Van Gils *et al.* [5]; Rosenblatt [6]). In opposition to this opinion, some researchers have found that family firms may withstand crisis exposure, because of having strong family and stakeholder relations (Sirmon *et al.* [7]; Corbetta and Salvato [8]; Anderson and Reeb [9]). These contradictory views call for further research of the behavior of family firms in crises.

The objective of this paper is to measure the performance of construction companies listed on the Warsaw Stock Exchange with respect to two portfolios: family- and non-family controlled. The performance of these portfolios was measures in three sub-periods: (1) pre-crisis period: 2006–2007; (2) crisis period: 2008–2009; (3) post-crisis period: 2010–2012. The capital asset pricing model was used to conclude whether family-controlled firms underperformed or outperformed their non-family peers in terms of expected returns and risk.

Table 1. Annual growth, index of production for construction, EU-27 and Euro area, 2000–2010 ([1]) (% growth; comparison with the previous year).

		2000	2001	2002	2003	2004	2005	2006	2007	2008	2009	2010
EU-27	Total construction	4.0	0.5	1.1	1.8	0.8	1.9	3.6	2.0	−3.8	−8.5	−4.1
	Building	5.2	0.4	0.7	2.7	1.8	2.2	4.3	1.9	−4.3	−10.9	−3.4
	Civil engineering	0.2	1.1	1.7	−1.5	−3.2	1.9	−1.2	2.9	−1.4	2.3	−7.1
EA-17	Total construction	4.1	0.6	0.4	0.9	0.0	2.2	3.7	1.3	−5.5	−7.9	−7.7
	Building	5.3	0.4	−0.1	1.6	0.8	2.6	4.7	1.2	−6.1	−10.1	−6.7
	Civil engineering	1.4	0.8	0.5	−0.8	−2.1	2.7	−1.8	2.1	−4.2	0.6	−12.0

([1]) Estimates; working day adjusted series. Source: Eurostat.

Table 2. Annual growth, index of production for total construction, 2000–2010 ([1]) (% growth; comparison with the previous year).

	Weight in 2005 (% of EU-27)	2000	2001	2002	2003	2004	2005	2006	2007	2008	2009	2010
EU-27	100.0	4.0	0.5	1.1	1.8	0.8	1.9	3.6	2.0	−3.8	−8.5	−4.1
EA-17	71.3	4.1	0.6	0.4	0.9	0.0	2.2	3.7	1.3	−5.5	−7.9	−7.7
Belgium	2.0	:	−2.4	−2.6	−0.2	2.9	0.6	3.3	1.5	−0.4	−3.3	−2.1
Bulgaria	0.1	:	13.4	3.8	4.6	35.3	31.7	24.8	26.8	12.6	−14.2	−17.9
Czech Republic	0.8	0.8	10.3	2.9	9.5	8.6	5.2	6.4	6.8	−0.3	−0.6	−7.3
Denmark	1.7	1.7	−6.7	−1.2	2.1	−0.2	3.1	3.8	−4.2	−5.7	−10.8	−8.4
Germany	11.1	−3.5	−7.6	−4.3	−4.2	−5.3	−5.3	6.3	2.9	−0.7	0.1	0.2
Estonia	0.1	18.6	4.2	22.6	6.1	12.5	22.4	26.9	13.5	−13.3	−29.8	−12.4
Ireland	2.8	:	3.4	2.0	5.7	25.3	10.0	3.8	−13.5	−29.2	−36.9	−30.1
Greece	1.1	:	6.6	39.1	−5.7	−15.9	−38.7	3.6	14.3	7.7	−17.5	−31.6
Spain	18.8	10.7	3.0	0.6	7.2	2.3	10.9	2.2	−4.3	−16.3	−11.3	−20.2
France	12.5	6.0	1.2	−2.3	−0.8	−1.2	2.7	4.2	2.3	−3.7	−5.9	−3.4
Italy	11.5	6.2	6.2	5.1	2.8	1.6	1.3	3.9	6.4	−1.1	−11.5	−3.4
Cyprus	0.3	:	3.7	3.2	6.5	4.4	2.9	4.1	6.8	2.3	−10.6	−8.0

Latvia	0.1	7.2	6.2	12.1	13.1	13.1	15.5	13.3	13.6	−3.1	−34.9	−23.4
Lithuania	0.2	−18.1	7.4	21.7	27.8	6.8	9.9	21.7	22.2	4.0	−48.5	−7.7
Luxembourg	0.3	:	4.2	1.9	0.9	−1.1	−0.9	2.6	2.6	−1.8	0.8	0.1
Hungary	0.5	7.7	9.3	18.0	2.7	4.3	15.7	−0.7	−14.0	−5.2	−4.4	−10.4
Malta	0.0	:	−4.0	23.4	−5.7	8.0	18.5	4.4	7.2	6.6	−7.9	0.2
Netherlands	4.9	:	1.9	−3.1	−4.9	−2.6	3.2	2.3	5.6	3.2	−3.0	−11.0
Austria	2.2	−0.6	−0.6	0.5	12.3	5.0	4.9	5.9	3.9	−0.9	−1.6	−4.3
Poland	1.7	1.2	−10.9	−10.1	−7.2	−1.9	9.2	15.6	16.3	10.2	4.5	3.7
Portugal	1.8	:	4.7	−1.1	−8.6	−4.4	−4.5	−6.3	−4.0	−1.2	−6.6	−8.5
Romania	0.4	:	11.5	4.5	3.2	1.4	6.6	15.6	33.1	26.7	−15.2	−13.4
Slovenia	0.2	2.9	−10.5	7.5	9.6	0.7	2.0	15.7	18.5	15.5	−20.9	−16.9
Slovakia	0.2	0.2	0.7	4.1	5.7	5.9	14.5	15.4	5.5	11.5	−11.2	−4.3
Finland	1.3	8.0	−0.1	1.4	4.2	4.4	5.2	7.8	10.2	4.1	−13.2	11.9
Sweden	2.1	4.4	5.4	0.3	0.0	0.1	3.0	8.0	6.2	4.2	−3.5	5.9
United Kingdom	21.2	4.2	1.1	4.6	5.6	3.5	−0.5	1.4	2.3	−1.3	−11.6	7.3
Norway	-	−2.1	1.2	−0.1	2.1	7.4	8.9	6.0	5.8	1.1	−8.3	−0.1
Switzerland	-	2.7	−2.7	0.9	0.1	3.1	2.6	2.0	1.2	2.4	1.4	1.9
Montenegro	-	:	5.3	0.3	−8.5	−5.8	7.6	46.1	−1.7	20.7	−19.3	−0.6
Croatia	-	−8.9	3.7	13.0	22.3	1.6	−0.2	9.3	2.6	11.8	−6.9	−15.9
FYR of Macedonia	-	:	:	:	:	:	:	−12.3	7.5	25.5	13.7	15.2
Turkey	-	:	:	:	:	:	:	18.4	5.5	−7.6	−16.3	17.5

(1) Estimates; working day adjusted series. Source: Eurostat.

Since the seminal findings of Sharpe [10], Lintner [11] and Mossin [12], CAPM has gained the immense interest of researchers. Some questioned, while others supported, the positive and linear relation between the risk and return of securities that CAPM predicted. In addition, over the past few decades, researchers have tried to relax CAPM assumptions and apply this model in a wide array of contexts. The cross-sectional behavior of asset returns has been researched voluminously in U.S., though with limited interest in Poland (Lyn and Zychowicz [13]; Kompa and Matuszewska [14]; Witkowska [15]; Gębka [16]; Kowalewski *et al.* [17]; or Waszczuk [18]).

Risk and return are two basic and important financial indicators for investors. In general, investors expect more returns for taking more risk. This linear relationship is reflected in the CAPM model and is still perceived as being powerful. Its components as beta or as systematic risk allow the prediction of rates on returns of securities portfolios.

This paper is structured as follows: The next section provides the theoretical background of CAPM with an outline of its evolution and supportive and contradictory views. In the methodological part, we present portfolio construction, address the public family firms and test data retrieved from the Warsaw Stock Exchange. The final section discusses the results and limitations.

2. THEORETICAL BACKGROUND

The capital asset pricing model was developed almost simultaneously by Sharpe [10], Lintner [11] and Mossin [12] based on Harry Markovitz's modern portfolio theory [19]. This model assumes the mean variance as the investment allocation predictor to one-period utility maximizing investors. Five decades later, this model and its variants are still widely used in finance to explain and predict the performance of asset portfolios, the cost of capital and the relation between asset risk and expected return, among others things. In particular, this theory assumes that investors should be rewarded by higher asset returns for taking higher risks. In essence, CAPM assumes two types of risk: systematic (non-diversifiable), which is related to the market, and specific and diversifiable (or unsystematic), which are related to individual securities that may be diversified. Unsystematic risk may be minimalized by effectively diversifying and increasing the portfolio size. On the other hand, systematic risk relates to the overall movement of the market or economy and cannot be diversified. As a consequence, CAPM helps investors in measuring a security risk with respect to the market. Testing CAPM over the long term, researchers have replicated this model in many contexts and countries and are still divided regarding its viability. Some find supportive and some contradictory evidence against CAPM.

Jensen *et al.* [20] proved the validity of CAPM on the New York Stock Exchange by testing linearity in a cross-section of expected monthly excess returns and betas on ten portfolios between 1931 and 1965. Fama and McBeth [21] found a positive linear relationship between average returns and the beta of assets on the NYSE between 1926 and 1968. Isakov [22], in his examination of the Swiss stock market between 1983 and 1991, proved the viability of CAPM. Zhang and Wihlborg [23] tested a monthly time series of equity prices by analyzing 753 firms quoted in six European emerging stock markets: Cyprus, Czech Republic, Greece, Hungary, Russia and Turkey between 1995 and 2002. They found a positive correlation between beta and returns for domestic CAPM. Litzenberger and Ramaswamy [24] found a positive relationship between dividend yield and stock return for the 1936–1977 period. Durack *et al.* [25] found supportive evidence for the CAPM model by testing Australian companies. Köseoğlu and Mercangöz [26] proved the validity of the zero beta capital asset pricing model by analyzing stocks within the ISE 100 on the Istanbul Stock Exchange.

However there is another stream of research against CAPM. Ibbotson and Sinquefield [27] revised the CAPM formula by ruling out beta and arguing that returns are better explained by the size of company and its book to market ratio. Roll [28] proved CAPM misspecification as the result of not finding an adequate measure of market portfolio. Banz [29] challenged CAPM by analyzing NYSE stocks between 1936 and 1975 and found that small firms had higher risk adjusted returns than larger firms. He argued that the size effect had existed for at least forty decades, which stood against the CAPM model. However, he argued that this effect was not linear and not stable; thus, the conclusions may not be supportive. Fama and French [30] analyzed a hundred portfolios on NYSE, AMEX and NASDAQ between July 1963, and December 1990. Their findings proved that size mattered, but not the relationship between beta and returns. Michailidis *et al.* [31] tested weekly stock returns of 100 firms listed in Greece between January 1998, and December 2002, and arrived at conclusions that higher risk is correlated with higher return, but the reverse correlation does not hold. Adedokun and Olakojo [32] arrived at the same results by testing monthly stock returns from a hundred Nigerian firms between January 2008, and December,2009. Hasan *et al.* [33] tested CAPM for the Bangladesh stock market and found out that this model does not prove its validity.

Mixed results of testing CAPM encouraged researchers to adjust this model by other variables than just using beta to explain returns. Merton [34] introduced intertemporal beta to differentiate investor and market beta. In the first case, investors bear greater risk (cash flow) than the market overall (discount beta equal to the variance of the market return), which should be reflected in higher returns. Breeden [35] proposed adjusting Merton's beta by the consumption aggregate in a single equation. Basu [36] argued that the relationship between price-earnings ratios is stronger than

CAPM predicts. Bhandari [37] found a positive relationship between debt to equity and expected stock returns. This relation holds true, even with different controlling variables, such as firm size, beta and the January effect. In addition, the relationship holds, even without the January effect, though it is weaker. Some researchers found that leverage and the book-to-market ratio better explain higher asset returns than CAPM (Rosenberg *et al.*, [38]; Chan *et al.* [39]). In addition, Chan *et al.* [39] measured the relation between stock returns and four variables: earnings yield, size, book to market ratio and cash flow yield in Japan. They found that all four of these variables correlate with market return, out of which, the book to market ratio and cash flow yield correlate the most with returns. In line with these vast contributions, Fama and French [40] revised their model by analyzing the two variables of size and book-to-market ratio in addition to excess market return. Javid and Ahmad [41] argued that conditional co-skewness explains non-normality in stock prices. Akbari and Mohammadi [42] found no relation between beta and the leverage ratios of 115 firms quoted at the Tehran Stock Market between 2005 and 2012.

There is very limited interest in research testing CAPM on family firms. Zellweger [43] pointed out the revision of CAPM to address the time horizon inherent in family firms. Zellweger argues that by extending the time horizon for investments, the marginal risk of an investment is reduced.

Some research studies have focused on testing CAPM on the Warsaw Stock Exchange. Lyn and Zychowicz [13] found book to market and dividend yield ratios in some Eastern European markets to be predictors of returns over six to 12 months. Kompa and Matuszewska [14] researched returns of five companies with the highest percentage shares listed under the Warsaw Stock Exchange Index 20 that characterize the largest and the most liquid companies. Witkowska [15] analyzed five portfolios composed of securities between 2002 and 2005 and found proof for the viability of the CAPM on the WSE. Gębka [16] found, for example, cross-autocorrelation between size and volume in portfolio returns during the 1996–2000 period. Waszczuk [18] showed that momentum trading may not be taken into account to generate profits beyond a one-month holding period between 2002 and 2011. In addition, she found proof for applying size and value strategies; however, their relevance varies over time. Almost none of the analyses of quoted Polish securities focused on measuring the performance of family firms. The only study that addressed family firms was performed by Kowalewski *et al.* [17]. They analyzed public family firms and found a relationship between the share of ownership and firm performance by analyzing 217 Polish companies from 1997 to 2005.

This study aims at contributing to the field of CAPM testing by analyzing securities of family-controlled *vs.* non-family peers, in particular measuring their performance in the crisis period.

3. METHODOLOGY

3.1. Definition

Chua *et al.* [44] analyzed the literature on the definition of family businesses, where they referred to Alcorn's definition of a public family firm as follows: "if part of the stock is publicly owned, the family must also operate the business". This rests in line with the principles of the Wallenbergs, the owners of the group of family businesses termed as the Wallenberg sphere. The Wallenbergs use the term "active ownership" to express family engagement in the businesses (Lindgren *et al.* [45]). This does not rely on quantification of ownership shares, but rather, controlling firms both in formal and informal ways. The minimum degree of active ownership means taking an active part in management tasks in the sense of influencing managerial and supervisory functions.

Jaskiewicz*et al.* [46] define family businesses by numbers, *i.e.*, having "family ownership of more than 25%, as well as family control and management participation". Blondel *et al.* [47] analyzed family firms quoted on the Partis Stock Exchange defined as: "a business where one or several individuals or families are identified as the major ultimate owner or owners". Leach and Bogod [48] classified a firm "as family controlled when at least 25% of shares are in the hands of family shareholders (provided the other 75% shareholding is distributed across smaller minority shareholders) and the business has experienced generational change". Villalonga and Amit [49] examine their research findings and assume a family firm with "a minimum control threshold of 20% of the votes, being the largest shareholder or voteholder, having family officers or directors, or being in second or later generation". Sraer and Thesmar [50] followed the definition of Villalonga and Amit [49] to measure the performance of public family firms in France. Miller *et al.* [51] stress the importance of family involvement in the ownership from the time perspective. They define "a family firm as one in which multiple members of the same family are involved as major owners or managers, either contemporaneously or over time". We follow the definition of Villalonga and Amit [49].

3.2. Data Description

We analyzed monthly weighted portfolio returns consisting of 24 public construction firms that belong to the Warsaw Stock Exchange Construction Index (WIG-Construction) (see Table 3 andTable 4).

Table 3. WIG-Construction portfolio in PLN (family-controlled firms in italics).

Company	Market Cap		Family	
	Market Cap	%	Shares %	Votes %
Budimex	1,303,762,250	36.62%	-	-
Elektrobudowa	498,435,000	14.00%	-	-
Trakcja PRKII	322,712,640	9.06%	-	-
Mostostal Zabrze	190,973,650	5.36%	-	-
Erbud	163,726,800	4.60%	-	-
Polimex-Mostostal	134,416,620	3.78%	-	-
Ulma-Construccion Polska	121,458,400	3.41%	-	-
Instal Kraków	117,900,450	3.31%	-	-
Unibep	*101,347,200*	*2.85%*	*26.98% ** *17.83%*	*26.98%* *17.83%*
P.A. Nova	*94,367,000*	*2.65%*	*18.19%* *9.13%*	*20.80%* *10.88%*
Prochem	66,588,900	1.87%	-	-
Mirbud	*63,918,020*	*1.80%*	*43.19%*	*43.19%;*
Herkules	*62,118,000*	*1.74%*	*18.99%*	*18.99%;* *93.31% ***
Elektrotim	61,625,740	1.73%	-	-
ZUE	*58,560,000*	*1.64%*	*72.75%* *0.01%*	*72.75%* *0.01%*
Projprzem	50,534,820	1.42%	-	-
Mostostal Warszawa	39,325,140	1.10%	-	-
Centrum Nowoczesnych Technologii	34,203,650	0.96%	-	-
Tesgas	*20,029,750*	*0.56%*	*40.58%*	*55.66%*
Mostostal Plock	14,261,400	0.40%	-	-
Energoaparatura	13,069,400	0.37%	-	-
Bipromet	12,009,900	0.34%	-	-
Mostostal Export	7,941,450	0.22%	-	-
Awbud	7,413,450	0.21%	-	-
Non-family cap	3,160,359,660	88.76%	-	-
Family-controlled cap	400,339,970	11.24%	-	-
TOTAL CAP	3,560,699,630	100.00%	-	-

* Two owning families; ** % in the Extraordinary General Meeting. Source: Warsaw Stock Exchange [52].

Table 4. WIG-Construction companies by expected returns and betas (not adjusted for size).

Company	2006–2007		2008–2009		2010–2011	
	Expected Return	**Beta**	**Expected Return**	**Beta**	**Expected Return**	**Beta**
Budimex	0.04	1.12	0.00	0.48	0.01	1.34
Elektrobudowa	0.08	1.23	−0.01	0.38	−0.02	0.88
Trakcja PRKII	-	-	−0.01	0.59	−0.06	2.12
Mostostal Zabrze	0.11	2.11	−0.01	1.31	−0.04	2.16
Erbud	0.03	0.51	0.00	1.16	−0.04	0.82
Polimex-Mostostal	0.02	1.45	−0.03	1.17	−0.03	1.50
Ulma-Construccion Polska	0.11	0.70	−0.02	1.40	−0.01	0.31
Instal Kraków	0.07	0.98	0.00	1.00	−0.01	0.85
Unibep	-	-	*0.30*	*1.19*	*0.01*	*0.48*
P.A. Nova	*0.01*	*0.31*	*0.28*	*0.71*	*0.01*	*1.04*
Prochem	0.02	1.17	−0.01	1.36	−0.01	1.64
Mirbud	-	-	*0.19*	*0.61*	*−0.01*	*0.59*
Herkules	*0.09*	*3.04*	*0.18*	*1.40*	*−0.06*	*1.40*
Elektrotim	−0.07	1.20	0.00	0.38	−0.02	1.39
ZUE	-	-	-	-	*−0.07*	*−0.05*
Projprzem	0.05	2.09	−0.03	0.69	−0.02	1.18
Mostostal Warszawa	0.09	0.73	0.01	0.52	−0.04	1.31
Centrum Nowoczesnych Technologii	0.18	3.30	0.00	1.40	0.02	0.18
Tesgas	-	-	*0.06*	*0.02*	*−0.03*	*0.78*
Mostostal Płock	0.08	1.69	−0.01	0.99	−0.04	1.81
Energoaparatura	0.05	1.90	−0.01	0.65	−0.03	1.21
Bipromet	-	-	−0.02	0.39	0.01	−0.09
Mostostal Export	0.08	3.59	−0.02	0.53	−0.01	1.36
Awbud	0.10	2.76	−0.04	0.58	0.01	0.74

Source: own computations based on the Warsaw Stock Exchange data [52].

Next, these firms were separated to form two sub-portfolios: family-controlled firms (6 firms) and nonfamily-controlled peers (18 firms). The firm size was used as weight in the portfolio formation. We calculated beta estimates and portfolio expected returns followed by Blume [53], Friend and Blume [54] and Black *et al.* [55]:

$$\beta_{constr} \equiv \sum_{i=1}^{N} x_{ip}\beta_i \,;\; E(R_{constr}) = \sum_{i=1}^{N} x_{iconstr} E(R_i) \tag{1}$$

where $x_{iconstr}$ denotes weights i = 1,2,..., N for assets in the construction portfolio (*constr*), $E(R_{constr})$ expected return on the construction portfolio and $E(R_i)$ expected return on security i.

The financial data for both of these types of portfolios were analyzed from 2006 to 2012 and divided into three sub-periods: (1) before the crisis: 2006–2007; (2) in crisis: 2008–2009; and (3) after the crisis 2010–2012 (see Table 5).

Table 5. Portfolio population.

	Before The Crisis	In Crisis	After The Crisis
Portfolio formation period	2006–2007	2008–2009	2010–2012
No. of family firms	2	5	6
No. of nonfamily firms	16	18	18

3.3. Methodology and Results

The CAPM model assumes a linear dependency between expected return on any asset with the expected return on the market portfolio. Jagannathan and McGratten [56] reflected that CAPM explains differences in the risk premium across assets, which helps investors make decisions. The risk is equivalent with beta. The slope of the CAPM line expresses the market risk premium, while the intercept represents the risk-free rate. Black *et al.* [55] and Fama and MacBeth [21] proposed a two-pass methodology to test the viability of CAPM. The first pass estimates portfolio betas, while the second pass regresses average returns based on the first-pass estimated portfolio betas. Sharpe [10], Lintner [11] and Mossin [12] provided the formula that explains the excess return of assets by risk premium:

$$E(R_i) = R_f + \beta_i\,(E(R_m) - R_f) \tag{2}$$

where $E(R_i)$ is the expected return on security i, R_f the return of the risk-free security, β_i the systematic risk of the security i calculated as $Cov(R_i,R_m)/Var(R_m)$ and $E(R_m)$ the expected return on the market portfolio.

Jensen [20] concludes that the correlation between beta and expected returns requires a time-series regression formulated as follows:

$$R_i - R_f = \alpha_i + \beta_i(R_m - R_f) + \varepsilon_i \tag{3}$$

Jensen's alpha defines the pricing error (abnormal performance). When alpha increases, the reliability of return estimates may deteriorate. Jensen's alpha shall equal zero to hold the CAPM model. We followed Formulas (4) and (5) to measure the performance of construction portfolios on the Warsaw Stock Exchange. We calculated the total risk as the variance of returns:

$$\sigma^2_{constr} = \sum_{i=1}^{N} x_{iconstr} + \left(\sum_{i=1}^{N} x_{iconstr}\beta_i \right)^2 \sigma_m{}^2 \tag{4}$$

and unsystematic risk as the transformation of the Equation (4):

$$\sigma_{\varepsilon i}{}^2 = \sigma_{constr}{}^2 - \beta_i^2\,\sigma_m{}^2 \tag{5}$$

We used the Warsaw Stock Exchange Index as the proxy for the construction market portfolio and the reference rates (minimum money market intervention rate) published by the National Bank of Poland as the proxy for the risk-free rate. Then, we tested stock returns for the normal distribution by using the Jarque–Bera test. The Jarque–Bera test relies on the third and fourth momentum and follows $\chi 2$ distribution with two degrees of freedom:

$$\textit{Jarque-Bera} = \frac{n}{6}\left[S^2 + \frac{(K-3)^2}{4}\right] \quad (6)$$

where S is the skewness, K the kurtosis and n the number of observations. The null hypothesis of returns following the normal distribution was tested at a p-value of 95%. The results are presented in Table 6.

The Jarque–Bera test proved that the normal distribution has to be rejected for the period from 2006 to 2007 ($\chi^2_{0.05;2} > 5.991$). However, the normal distribution test may not be rejected for all remaining portfolios between 2008 and 2012 ($\chi^2_{0.05;2} \leq 5.991$), and therefore, the CAPM model may be used to explain the relationship between the market returns and risk of construction portfolios. Before the crisis, both portfolios reported positive returns, and non-family firms outperformed family-controlled ones by 1.22%. The returns were opposite in the crisis period, when family firms significantly outperformed non-family peers. In addition, the latter firms reported negative returns in the crisis. The data shows that tolerance for risk has declined since the crisis hit. This behavior seems to be rational, as invoked by the reaction to the global recession. In addition, public construction companies exhibited high proportions of specific, rather than systematic, risk. In periods before and after the crisis, all construction portfolios were oversensitive with respect to the market ($\beta > 1$). During the crisis, both portfolios were less sensitive to the market ($\beta < 1$). When the crisis hit, all public construction companies were overvalued, except for family-controlled ones, which were undervalued.

Table 6. Tests of two construction portfolios on the Warsaw Stock Exchange (WSE): non-family controlled firms (NFF) and family-controlled firms (FF).

	2006–2007		2008–2009		2010–2012	
	NFF	FF	NFF	FF	NFF	FF
Expected Return	5.35%	4.13%	−0.76%	23.56%	−1.56%	−1.72%
Variance	0.093	0.071	0.022	0.031	0.011	0.010
Standard Deviation	0.306	0.267	0.150	0.175	0.107	0.103
Coefficient of variation	5.719	6.467	−19.816	0.742	−6.816	−5.963
Beta	1.223	1.392	0.680	0.921	1.305	0.711
Systematic Risk	0.005	0.006	0.004	0.007	0.004	0.001
Specific Risk	0.089	0.065	0.018	0.023	0.008	0.009
Specific Risk/Variance	0.948	0.912	0.819	0.757	0.661	0.892
Jensen's alpha	0.032	0.018	−0.008	0.237	−0.018	−0.020
Skewness	6.142	1.231	0.438	−0.623	0.175	0.212
Kurtosis	72.109	0.471	2.210	5.022	2.206	1.722
Jarque–Bera	3284.634	1.039	1.043	1.175	0.564	0.453

4. CONCLUSIONS

This research focused on testing the validity of CAPM on the Warsaw Stock Exchange across two portfolios of family- and non-family-controlled companies. These portfolios were constructed on the basis of the complete data retrieved from the Warsaw Stock Exchange. As a consequence, we made calculations on the whole population (not sample) data. In this respect, even the small amount of data does not exclude an inference. These two portfolios were then adjusted for the size,*i.e.*, market capitalization, to avoid the dominance of large companies in the population.

The assumption of normal distribution may not be rejected according to the Jarque–Bera test for the period from 2008 to 2012; therefore, the relation between return and risk holds in the case of the WSE. However, for the period from 2006 to 2007, the Jarque–Bera yielded outlying numbers; therefore, we did not draw conclusions.

By assuming that market portfolios were efficient, returns of family construction companies significantly outperformed their nonfamily peers in the latest crisis. Investors who invested in family-controlled portfolios during that time might have expected exceptional returns of 23.56%. In addition, these securities were undervalued, which does not prove CAPM invalidity. As Jensen argued [55], "deficiencies as measurement and model specification error due to proxies for variables do not reject CAPM".

Although the capital asset pricing model was substantially developed since its inception (see Theoretical Background), we decided to prove its validity as developed by Sharpe [10], Lintner [11] and Mossin [12], because of scant testing in the case of family firms. However it would be tempting to move beyond this version of CAPM to get more insight into the performance of family firms on stock exchanges.

This research was limited to public companies. In addition to the extraordinary returns of family-controlled firms, the results suggest that a vast amount of risk is specific for all construction portfolios. It would be interesting to find out the sources of this risk. As mentioned in the Introduction, Poland was the only EU country in which the construction industry withstood the crisis and reported positive returns. However, as this research is limited to public companies, it would be interesting to analyze in-depth factors that allowed family-controlled firms operating in the construction sector to outperform their non-family peers. In addition, this paper used the capital asset pricing model to explain the performance of two portfolios that used a family-ownership as the distinction factor. The outperformance of the family-owned firms during the crisis may be either attributed to family-ownership or non-family-related factors and being the result of portfolio composition. This issue also needs to be further analyzed, *i.e.*, by conducting a qualitative research.

CONFLICTS OF INTEREST

The author declares no conflict of interest.

REFERENCES

1. International Labour Office, Governing Body. *The Current Global Economic Crisis: Sectoral Aspects*; International Labour Office, Governing Body: Geneva, Switzerland, 2009; pp. 1–13.

2. Organisation for Economic Co-operation and Development (OECD). Responding to the Economic Crisis. Fostering

Industrial Restructuring and Renewal. Available online: http://www.oecd.org/ industry/ind/43387209.pdf (accessed on 30 January 2014).

3. Current Population Survey (CPS). United States Census Bureau. Available online: https://www.census.gov/cps/ (accessed on 30 January 2014).

4. Lee, J. Family firm performance: Further evidence. *Fam. Bus. Rev.* **2006**, *19*, 103–114.

5. Van Gils, A.; Voordeckers, W.; van den Heuvel, J. Environmental uncertainty and strategic behavior in Belgian family firms. *Eur. Manag. J.* **2004**, *22*, 588–595.

6. Rosenblatt, P.C. The interplay of family system and business system in family farms during economic recession. *Fam. Bus. Rev.* **1991**, *4*, 45–57.

7. Sirmon, D.; Arregle, J.; Hitt, M.; Webb, J. The role of family influence in firms' strategic responses to threat of imitation. *Entrep. Theory Pract.* **2008**, *32*, 979–998.

8. Corbetta, G.; Salvato, C. Self-serving or self-actualizing? Models of man and agency costs in different types of family firms: A commentary on comparing the agency costs of family and non-family firms: Conceptual issues and exploratory evidence. *Entrep. Theory Pract.* **2004**, *28*, 355–362.

9. Anderson, R.; Reeb, D.M. Founding-Family Ownership and Firm Performance: Evidence from the S&P 500. *J. Financ.* **2003**, *58*, 1301–1327.

10. Sharpe, W.F. Capital Asset Prices: A Theory of Market Equilibrium under conditions of risk.*J. Financ.* **1964**, *19*, 425–442.

11. Lintner, J. The Valuation of Risk Assets and the Selection of Risky Investments in Stock Portfolios and Capital Budgets. *Rev. Econ. Stat.* **1965**, *47*, 13–25.

12. Mossin, J. Equilibrium in a capital Asset Market. *Econometrica* **1966**, *34*, 768–783.

13. Lyn, E.O.; Zychowicz, E.J. Predicting Stock Returns in the Developing Markets of Eastern Europe. *J. Invest.* **2004**, *13*, 63–71.

14. Kompa, K.; Matuszewska, A.M. Rates of Return Analysis: Examples from the Warsaw Stock Exchange. *Int. Adv. Econ. Res.* **2007**, *13*, 111–112.

15. Witkowska, D.M. Portfolio Analysis: Example for the Warsaw Stock Exchange. *Int. Adv. Econ. Res.* **2007**, *13*, 247–248.

16. Gębka, B. Volume- and size-related lead–lag effects in stock returns and volatility: An empirical investigation of the Warsaw Stock Exchange. *Int. Rev. Financ. Anal.* **2008**, *17*, 134–155.

17. Kowalewski, O.; Talavera, O.; Stetsyuk, I. Influence of Family Involvement in Management and Ownership on Firm Performance: Evidence from Poland. *Fam. Bus. Rev.* **2010**, *23*, 23–45.

18. Waszczuk, A. Do Local or Global Risk Factors Explain the Size, Value and Momentum Trading Payoffs on the Warsaw Stock Exchange? *Appl. Financ. Econ.* **2013**, *23*, 1497–1508.

19. Markowitz, H.M. Portfolio Selection. *J. Financ.* **1952**, *7*, 77–91.

20. Jensen, M.C. The performance of mutual funds in the period 1945–1964. *J. Financ.* **1968**, *23*, 389–416.

21. Fama, E.F.; MacBeth, J.D. Risk, Return, and Equilibrium: Empirical Tests. *J. Polit. Econ.* **1973**,*81*, 607–636.

22. Isakov, D. Is beta still alive? Conclusive evidence from the Swiss stock market. *Eur. J. Financ.***1999**, *5*, 202–212.

23. Zhang, J.; Wihlborg, C. Unconditional and conditional CAPM: Evidence from European emerging markets. Available online: http://snee.org/filer/papers/266.pdf (accessed on 30 January 2014).

24. Litzenberger, R.H.; Ramaswamy, K. The Effects of Dividends on Common Stock Prices Tax Effects or Information Effects? *J. Financ.* **1982**, *37*, 429–443.

25. Durack, N.; Durand, R.B.; Maller, R.A. A best choice among asset pricing models? The Conditional Capital Asset Pricing Model in Australia. *Account. Financ.* **2004**, *44*, 139–162.

26. Köseoğlu, S.D.; Mercangöz, B.A. Testing the Validity of Standard and Zero Beta Capital Asset Pricing Model in Istanbul Stock Exchange. *Int. J. Bus. Humanit. Technol.* **2013**, *3*, 58–67.

27. Ibbotson, R.; Sinquefield, R. Stocks, Bonds, Bills and Inflation: Year-by-Year Historical Returns (1926–1974). *J. Bus.* **1976**, *49*, 11–47.

28. Roll, R. A Critique of the Asset Pricing Theory's Tests. *J. Financ. Econ.* **1977**, *4*, 129–176.

29. Banz, R.W. The relationship between return and market value of common stocks. *J. Financ. Econ.* **1981**, *9*, 3–18.

30. Fama, E.F.; French, K.R. The Cross-Section of Expected Stock Returns. *J. Financ.* **1992**, *47*, 427–465.

31. Michailidis, G.; Tsopoglou, S.; Papanastasiou, D.; Mariola, E. Testing the Capital Asset Pricing Model (CAPM): The Case of the Emerging Greek Securities Market. *Int. Res. J. Financ. Econ.* **2006**, *4*, 78–91.

32. Adedokun, A.J.; Olakojo, S.A. Test of Capital Asset Pricing Model: Evidence from Nigerian Stock Exchange. *J. Econ. Theory* **2012**, *6*, 121–127.

33. Hasan, Z.; Kamil, A.A.; Baten, A. Analyzing and estimating portfolio performance of Bangladesh stock market. *Am. J. Appl. Sci.* **2013**, *10*, 139–146.

34. Merton, R.C. An intertemporal capital asset pricing model. *Econometrica* **1973**, *41*, 867–887.

35. Breeden, D.T. An intertemporal asset pricing model with stochastic consumption and investment opportunities. *J. Financ. Econ.* **1979**, *7*, 265–296.

36. Basu, S. Investment Performance of Common Stocks in Relation to Their Price-Earnings Ratios: A Test of the Efficient Market Hypothesis. *J. Financ.* **1977**, *32*, 663–682.

37. Bhandari, L. Debt/Equity Ratio and Expected Common Stock Returns: Empirical Evidence. *J. Financ.* **1988**, *2*, 507–528.

38. Rosenberg, B.; Reid, K.; Lanstein, R. Persuasive evidence of market inefficiency. *J. Portf. Manag.* **1985**, *11*, 9–16.

39. Chan, L.; Hamao, Y.; Lakonishok, J. Fundamentals and Stock Returns in Japan. *J. Financ.*.**1991**, *46*, 1739–1764.

40. Fama, E.F.; French, K.R. Common Risk Factors in the Returns on Stocks and Bonds. *J. Financ. Econ.* **1993**, *33*, 3–56.

41. Javid, A.Y.; Ahmad, E. *Test of Multi-moment Capital Asset Pricing Model: Evidence from Karachi Stock Exchange*; Pakistan Institute of Development Economics Working Paper 2008:49; Pakistan Institute of Development Economics: Islamabad, Pakistan, 2008.

42. Akbari, P.; Mohammadi, E. A Study of the Effects of Leverages Ratio on Systematic Risk based on the Capital Asset Pricing Model Among Accepted Companies in Tehran Stock Market. *J. Educ. Manag. Stud.* **2013**, *3*, 271–277.

43. Zellweger, T. Time Horizon, Costs of Equity Capital, and Generic Investment Strategies of Firms. *Fam. Bus. Rev.* **2007**, *20*, 1–15.

44. Chua, J.H.; Chrisman, J.J.; Sharma, P. Defining the Family Business by Behavior. *Entrep. Theory Pract.* **1999**, *23*, 19–39.

45. Lindgren, H.; Fritz, S. Aktivt Ägande. Investor under Växlande Konjunkturer. In*Scandinavian Economic History Review*; Routledge: Abingdon, UK, 1996. (in Swedish)

46. Jaskiewicz, P.; Gonzalez, V.M.; Menendez, S.; Schiereck, D. Long-run IPO Performance Analysis of German and Spanish Family-owned businesses. *Fam. Bus. Rev.* **2005**, *18*, 179–202.

47. Blondel, C.; Rowell, N.; van der Heyden, L. Prevalence of Patrimonial Firms on Paris Stock Exchange: Analysis of the Top 250 Companies in 1993 and 1998. Available online: http://www.insead.edu/facultyresearch/research/doc.cfm?did=990 (accessed on 30 January 2014).

48. Leach, P.; Bogod, T. *The BDO-Stoy Hayward Guide to the Family Business*; Kogan Page: London, UK, 1999.

49. Villalonga, B.; Amit, R. How do family ownership, control and management affect firm value? *J. Financ. Econ.* **2006**, *80*, 385–417.

50. Sraer, D.; Thesmar, D. Performance and Behavior of Family Firms: Evidence from the French Stock Market. Available online: http://papers.ssrn.com/sol3/papers.cfm?abstract_id=925415(accessed on 20 December 2013).

51. Miller, D.; Le Breton-Miller, I.; Lester, R.H.; Cannella, A.A., Jr. Are Family Firms Really Superior Performers. *J. Corp. Financ.* **2007**, *5*, 829–858.

52. Warsaw Stock Exchange. Available online: http://www.gpw.pl (accessed on 20 December 2013).

53. Blume, M.E. Portfolio Theory: A Step Toward Its Practical Application. *J. Bus.* **1970**, *43*, 152–173.

54. Friend, I.; Blume, M. Measurement of Portfolio Performance under Uncertainty. *Am. Econ. Rev.* **1970**, *60*, 607–636.

55. Black, F.; Jensen, M.C.; Scholes, M. The Capital Asset Pricing Model: Some Empirical Tests. In *Studies in the Theory of Capital Markets*; Jensen, M.C., Ed.; Praeger: New York, NY, USA, 1972; pp. 79–121.

56. Jagannathan, R.; McGrattan, E.R. The CAPM debate. *Fed. Reserv. Bank Minneap. Q. Rev.* **1995**,*19*, 2–17.

CHAPTER 11

Polynomial Regressions and Nonsense Inference

Daniel Ventosa-Santaulària [1,*]
and Carlos Vladimir Rodríguez-Caballero [2]

[1] Centro de Investigación y Docencia Económicas (CIDE), División de Economía, Carretera México-Toluca 3655 Col. Lomas de Santa Fe, Delegación Álvaro Obregón, México 01210, Mexico

[2] Center for Research in Econometric Analysis of Time Series (CREATES) and Department of Economics and Business, Aarhus University, Fuglesangs Allé 4, Building 2622 (203), Aarhus V 8210, Denmark

ABSTRACT

Polynomial specifications are widely used, not only in applied economics, but also in epidemiology, physics, political analysis and psychology, just to mention a few examples. In many cases, the data employed to estimate such specifications are time series that may exhibit stochastic nonstationary behavior. We extend Phillips' results (Phillips, P. Understanding spurious regressions in econometrics. *J. Econom.* **1986**, *33*, 311–340.) by proving that an inference drawn from polynomial specifications, under stochastic nonstationarity, is misleading unless the variables cointegrate. We use a generalized polynomial specification as a vehicle to study its asymptotic and finite-sample properties. Our results, therefore, lead to a call to be cautious whenever practitioners estimate polynomial regressions.

KEYWORDS

polynomial regression; misleading inference; integrated processes

1. INTRODUCTION

There is some research on the effects of the nonstationarity of the variables on nonlinear relationships (spurious inference on linear regressions was uncovered by [1], and later explained by [2]). In [3], it is shown (both in finite samples and asymptotically) that six nonlinear tests (the Ramsey Regression Equation Specification Error Test (RESET), McLeod and Li test, Keenan test, Neural Network test, White's information matrix, and the one proposed by [4]), when applied to independent random walks, tend to identify spurious (non-existing) nonlinear relationships (it is noteworthy that [5] studied the spurious regression phenomenon under stochastic nonstationarity when the logarithms of independent integrated of order one, I(1), variables are used; logarithmic transformations are commonly used in applied studies to deal with nonlinearity). The author of [6] extend these results by studying the behavior of two additional tests: the Brock-Dechert-Scheinkman (BDS) test and another one proposed by [7]; he finds that the former also yields results that do not make sense, whilst the latter proves to have good power properties even in small samples. The author of [8] studies the properties of the nonparametric Phillip's unit root test applied to polynomials of integrated processes and concludes, broadly speaking, that the tests does not possess an asymptotic nuisance-parameter-free distribution, except under very specific conditions.

To the best of our knowledge, the "nonlinear relationship-spurious inference" literature (briefly sketched earlier) focuses on statistical tests rather than polynomial regressions. The latter are used to linearly relate the dependent variable to a kthth order polynomial on an independent variable, x. Such regressions therefore fit (through ordinary least squares, OLS) a nonlinear relationship between a polynomial on the independent variable and the conditional mean of y.

These specifications can be traced back to the nineteenth century, to impute series (see [9]). Despite its old age, polynomial regressions remain widely used in a large number of scientific fields, which include epidemiology/disease progression [10], geophysics [11], physics [12], political analysis [13], psychology [14], and, of course, statistics. Splines regression models (cubic splines, for example) can be used to smooth/impute series.

In empirical economics, polynomial specifications can be found in many subfields, such as, financial economics [15,16], labor economics [17,18], agricultural economics [19], macroeconomics (exchange rates, [20]) and environmental economics [21,22]. An evocative example can be found in the empirical research dealing with the Kuznets curve and the environmental Kuznets curve; the inverse U-shaped relationship between the variables is typically specified as the dependent variable regressed on the independent and its square (see [23,24]; it is noteworthy that Kuznets' specifications usually employ even-order polynomials).

Even though polynomial regressions remain an important empirical tool, we could not find in the literature any attempt to study their properties when the variables behave as independent nonstationary processes. This might be so because the effect of nonstationarity is rather intuitive, and econometricians, at least those familiar with the spurious regression, could speculate that *t*-ratios diverge and the R2R2 does not collapse. However, many researchers in diverse fields seem to be unaware of this possibility.

In this paper, we confirm that an inference drawn from a polynomial regression, when the variables are generated as independent integrated processes, is misleading (when the variables cointegrate, inference drawn from such a specification is no longer misleading). We provide evidence that generalizes Phillip's results in two new directions: (i) we allow the exponent of the variables, both explanatory and dependent in a bivariate regression, to take any natural number; (ii) we allow for an arbitrary (natural number) order for the polynomial in x in a k-variate regression. The main objective of this work is to warn practitioners about the considerable risks of spurious inference when the powers of a nonstationary variable are used as regressors.

This paper is organized in a very simple manner. The next section presents the data-generating processes (DGPs) and the main results, divided in two theorems. A small Monte Carlo shows that the asymptotics are a sufficiently accurate representation of the finite sample behavior of the regressions.

2. ASYMPTOTICS OF POLYNOMIAL REGRESSIONS

The variables, both dependent and independent, are generated as independent driftless unit roots:

$$z_t = z_{t-1} + u_{z,t} \tag{1}$$

for z=x,y. The innovations, uy, tuy, t and ux, tux, t, are independent of each other and obey the conditions stated by Phillips ([2], p. 313, Assumption 1). We use these variables to estimate the following specification:

$$y_t^m = \alpha + \beta x_t^k + u_t \tag{2}$$

where m,k∈N. A word on notation; the symbol, →D→D, denotes weak convergence, and, for simplicity, Wz≡Wz(r), for z=x,y, denotes a Wiener standard process. The stochastic integral, ∫10, is written as ∫.

Theorem 1.

Let $\{y_t\}_{t=1}^{\infty}$ and $\{x_t\}_{t=1}^{\infty}$be independently generated by Equation (1). Estimate by OLS specification Equation (2). Then, as T→∞T→∞:

1. $T^{-\frac{m}{2}}\hat{\alpha} \xrightarrow{D} \sigma_y^m \left[\frac{\int w_y^m \int w_x^{2k} - \int w_x^k w_y^m \int w_x^k}{\int w_x^{2k} - \left(\int w_x^k\right)^2}\right]$

2. $T^{-\frac{1}{2}(m-k)}\hat{\beta} \xrightarrow{D} \frac{\sigma_y^m}{\sigma_x^k} \left[\frac{\int w_x^k w_y^m - \int w_x^k \int w_y^m}{\int w_x^{2k} - \left(\int w_x^k\right)^2}\right] \equiv \frac{\sigma_y^m}{\sigma_x^k}\widetilde{\beta}$

3. $T^{-\frac{1}{2}} t_{\hat{\beta}} \xrightarrow{D} \frac{\int w_x^k w_y^m - \int w_x^k \int w_y^m}{\left[\left(\int w_x^{2k} - \left(\int w_x^k\right)^2\right)\left(\int w_y^{2m} - \left(\int w_y^m\right)^2 - \left(\int w_x^k w_y^m - \int w_x^k \int w_y^m\right)^2\right)\right]^{\frac{1}{2}}}$

4. $R^2 \xrightarrow{D} \widetilde{\beta}^2 \left[\frac{\int w_x^{2k} - \left(\int w_x^k\right)^2}{\int w_y^{2m} - \left(\int w_y^m\right)^2}\right]$

Note that all these results are an extension of [2]. It is noteworthy to mention that, for k=m=1, our results are exactly those of [2]. This implies that, no matter what power does the practitioner applies to the variables, the spurious regression phenomenon remains identical. That said, a more interesting specification should allow for a more complete polynomial of the independent variable, as in:

$$y_t = \beta_0 + \beta_1 x_t + \beta_2 x_t^2 + \cdots + \beta_k x_t^k + u_t \quad (3)$$

where k∈N. In this case, OLS estimates still generate a spurious regression:

Theorem 2.

Let $\{y_t\}_{t=1}^{\infty}$ and $\{x_t\}_{t=1}^{\infty}$ be independently generated by Equation (1). Estimate by OLS specification Equation (3). Then, as T→∞:

Proof:

$$\begin{bmatrix} T^{-\frac{1}{2}}\hat{\beta}_0 \\ \hat{\beta}_1 \\ T^{\frac{1}{2}}\hat{\beta}_2 \\ \vdots \\ T^{\frac{1}{2}(k-1)}\hat{\beta}_k \end{bmatrix} \xrightarrow{D} \sigma_y \begin{pmatrix} 1 & 0 & 0 & \cdots & 0 \\ 0 & \sigma_x & 0 & \cdots & 0 \\ 0 & 0 & \sigma_x^2 & \cdots & 0 \\ \vdots & \vdots & \vdots & \cdots & \vdots \\ 0 & 0 & 0 & \cdots & \sigma_x^k \end{pmatrix}^{-1} \times$$

$$\begin{pmatrix} 1 & \int w_x & \int \omega_x^2 & \cdots & \int w_x^k \\ \int w_x & \int \omega_x^2 & \int \omega_x^3 & \cdots & \int \omega_x^{k+1} \\ \int \omega_x^2 & \int \omega_x^3 & \cdots & \cdots & \int \omega_x^{k+2} \\ \vdots & \cdot^{\cdot^{\cdot}} & \cdot^{\cdot^{\cdot}} & \cdot^{\cdot^{\cdot}} & \vdots \\ \int w_x^k & \cdots & \cdots & \cdots & \int \omega_x^{2k} \end{pmatrix}^{-1} \begin{pmatrix} \int w_y \\ \int \omega_x \omega_y \\ \int \omega_x^2 \omega_y \\ \vdots \\ \int \omega_x^k \omega_y \end{pmatrix}$$

1)

2) $S^2 = Op(T)$, where $S^2 = T^{-1}\sum_{t=1}^{T}\left(y_t - \sum_{i=0}^{k}\hat{\beta}_i x_t^i\right)^2$

3) $t_{\hat{\beta}_i} = Op\left(T^{\frac{1}{2}}\right)$ for $i = 0, 1, 2, \ldots, k$

Table 1. Rejection rates of *t*-ratios.

T	Specification (2)				Specification (3) With k = 4			
	k	m = 1	m = 2	m = 3	β_1	β_2	β_3	β_4
	1	0.77	0.71	0.71				
100	2	0.71	0.66	0.65	0.46	0.35	0.33	0.31
	3	0.72	0.66	0.66				
	1	0.85	0.82	0.82				
250	2	0.81	0.78	0.78	0.64	0.56	0.52	0.50
	3	0.82	0.78	0.78				
	1	0.89	0.87	0.87				
500	2	0.86	0.84	0.84	0.73	0.67	0.64	0.63
	3	0.88	0.84	0.84				

Note the linear pattern in the order of convergence of the parameters; whilst the constant term, β^0, diverges at rate T12, β^1 neither diverges, nor collapses, β^2 collapses at rate T−12T-12, and so on. Nonetheless, all the *t*-ratios of the estimated parameters diverge at the usual rate T12.

In both theorems, the convergence rate of the *t*-ratios associated with the estimates diverge. This implies that, for a sufficiently large sample, the null hypothesis that the parameters are equal to zero will eventually be rejected. Finite sample evidence suggests that this actually occurs in even rather small samples of 100–500 observations (Table 1).

3. CONCLUDING REMARKS

In this paper, we extended the results of what is known as spurious inference by studying the asymptotic and finite-sample behavior of the *t*-ratios in an OLS-estimated regression, where the dependent variable and/or the explanatory variable are nonlinearly transformed by means of a polynomial. When the variables are independent and stochastically nonstationary, the inference based on OLS estimates is misleading. Our results concern pure integrated of order one processes, but provide a natural guide to future research; near-integration, integrated of order two, and broken linear trend processes should be further studied. This result should be understood as a call to be cautious whenever practitioners estimate polynomial regressions.

ACKNOWLEDGMENTS

The first draft of the article was written while Carlos Vladimir Rodríguez-Caballero was visiting the Center for Research and Teaching in Economics (CIDE). He gratefully acknowledges Alejandro López-Feldman for his support.

CONFLICTS OF INTEREST

The authors declare no conflict of interest.

REFERENCES

1. Granger, C.; Newbold, P. Spurious regressions in econometrics. *J. Econom.* **1974**, *2*, 111–120.

2. Phillips, P. Understanding spurious regressions in econometrics. *J. Econom.* **1986**, *33*, 311–340.

3. Lee, Y.-S.; Tae-Hwan, K.; Newbold, P. Spurious nonlinear regressions in econometrics. *Econ. Lett.* **2005**, *87*, 301–306.

4. Hamilton, J.D. A parametric approach to flexible nonlinear inference. *Econometrica* **2001**, *69*, 537–573.

5. De Jong, R.M. Logarithmic spurious regressions. *Econ. Lett.* **2003**, *81*, 13–21.

6. O'Brien, E. A note on spurious nonlinear regression. *Econ. Lett.* **2008**, *3*, 366–368.

7. Peña, D.; Rodriguez, J. Detecting nonlinearity in time series by model selection criteria. *Int. J. Forecast.* **2005**, *21*, 731–748.

8. Wagner, M. The Phillips unit root tests for polynomials of integrated processes. *Econ. Lett.***2012**, *114*, 299–303.

9. Gergonne, J. The application of the method of least squares to the interpolation of sequences.*Hist. Math.* **1815**, *1*, 439–447, (Translated by Ralph St. John and S. M. Stigler from the 1815 French edition).

10. Chatterjee, C.; Sarkar, R. Multi-step polynomial regression method to model and forecast malaria incidence. *PLoS One* **2009**, *3*, e4726.

11. Verma, S. Evaluation of polynomial regression models for the Student *t* and Fisher *F* critical values, the best interpolation equations from double and triple natural logarithm

transformation of degrees of freedom up to 1000, and their applications to quality control in science and engineering. *Revista Mex. Ciencias Geol.* **2009**, *26*, 79–92.

12. Barker, P.; Street-Perrott, F.; Leng, M.; Greenwood, P.; Swain, D.; Perrott, R.; Telford, R.; Ficken, K. A 14,000-year oxygen isotope record from diatom silica in two alpine lakes on Mt. Kenya. *Science* **2001**, *292*, 2307–2310, p. 2,310.

13. Green, D.; Leong, T.; Kern, H.; Gerber, A.; Larimer, C. Testing the accuracy of regression discontinuity analysis using experimental benchmarks. *Polit. Anal.* **2009**, *17*, 400–417.

14. Shanock, L.; Baran, B.; Gentry, W.; Pattison, S.; Heggestad, E. Polynomial regression with response surface analysis: A powerful approach for examining moderation and overcoming limitations of difference scores. *J. Bus. Psychol.* **2010**, *25*, 543–554.

15. Ferrer, R.; González, C.; Soto, G. Linear and nonlinear interest rate exposure in Spain.*Manag. Financ.* **2010**, *36*, 431–451.

16. Ioannidis, C.; Peel, D.; Peel, M. The time series properties of financial ratios: Lev revisited. *J. Bus. Financ. Account.* **2003**, *30*, 699–714.

17. Leonardi, M.; Pica, G. Who pays for it? The heterogeneous wage effects of employment protection legislation. *Econ. J.* **2013**.

18. Straka, J. Is poor worker morale costly to firms? *Ind. Labor Relat. Rev.* **1993**, 381–394.

19. Ackello-Ogutu, C.; Paris, Q.; Williams, W. Testing a von Liebig crop response function against polynomial specifications. *Am. J. Agric. Econ.* **1985**, *67*, 873–880.

20. Darvas, Z. Estimation Bias and inference in overlapping autoregressions: Implications for the target-zone literature. *Oxf. Bull. Econ. Stat.* **2008**, *70*, 1–22.

21. Auffhammer, M.; Kellogg, R. Clearing the air? The effects of gasoline content regulation on air quality. *Am. Econ. Rev.* **2011**, *101*, 2687–2722.

22. Kellenberg, D. Trading wastes. *J. Environ. Econ. Manag.* **2012**, *61*, 68–87.

23. Grossman, G.; Krueger, A. Environmental Impacts of a North American Free Trade Agreement. In *The Mexico-U.S. Free Trade Agreement*; Garbe, P.M., Ed.; MIT Press: Cambridge, MA, USA, 1993; pp. 13–56.

24. Labson, S.; Crompton, P. Common trends in economic activity and metals demand: Cointegration and the intensity of use debate. *J. Environ. Econ. Manag.* **1993**, *25*, 147–161.

25. Giles, D. Spurious regressions with time-series data: Further asymptotic results. *Commun. Stat. Theory Methods* **2007**, *36*, 967–979.

26. Hannan, E.; Deistler, M. *The Statistical Theory of Linear Systems*; Cambridge University Press: Cambridge, MA, USA, 2012.

27. Golub, G.; van Loan, C. *Matrix Computations*; The Johns Hopkins University Press: Baltimore, MD, USA, 2013.

28. Trench, W. An algorithm for the inversion of finite Hankel matrices. *J. Soc. Ind. Appl. Math.* **1965**, *13*, 1102–1107.

29. Kumar, K.; Alsaleh, M. Application of Hankel matrices in polynomial regression. *Appl. Math. Comput.* **1996**, *77*, 205–211.

30. Hamilton, J.D. *Time Series Analysis*; Cambridge University Press: Cambridge, MA, USA, 1994.

Index